CARPENTRY AND BUILDING CONSTRUCTION STUDENT WORKBOOK

Fourth Edition

by John L. Feirer

Keyed to the 1993 edition of the textbook
Carpentry and Building Construction

GLENCOE

Macmillan/McGraw-Hill

New York, New York
Columbus, Ohio
Mission Hills, California
Peoria, Illinois

Send all inquiries to:
GLENCOE DIVISION
Macmillan/McGraw-Hill
3008 W. Willow Knolls Drive
Peoria, IL 61614

ISBN 0-02-668281-8

Printed in the United States of America

3 4 5 6 7 8 9 COU 98 97 96 95 94

TABLE OF CONTENTS

TO THE TEACHER

This student workbook has been designed for use with the textbook *Carpentry and Building Construction*. It has been planned as a teaching and learning aid. The exercises it presents are meant to reinforce and evaluate learning.

This student workbook can be used by students individually or in group activity. Pages are perforated, allowing individual study units to be removed for ease in use and grading.

Students can be asked to complete the exercises in this student workbook in one of several ways:

1. Each student may progress at his or her individual rate of speed, completing the study unit exercises as quickly as possible.
2. The teacher may assign a unit or section of a unit in the textbook as homework. The appropriate study unit may then be assigned the next day to evaluate learning.
3. Students can be assigned sections of the textbook for outside study. They may then be asked, at the completion of their study assignment, to complete the appropriate study units in the student workbook.

A list of the possible scores in study unit exercises is included. This list will enable students to track their progress.

Note that the questions in each study unit in this workbook relate to a particular text unit or units. The unit or units to which the study unit refers is given at the top left-hand corner of the first page of each study unit. The textbook page numbers of the unit or units are given immediately below this text unit reference. Note that the student work study units do not carry the same numbers as the textbook units.

Other material has been included in this student workbook. This additional material will simplify the teaching and be of value in:

- Organizing the class.
- Keeping accurate records of the cost of carpentry materials.
- Maintaining an accurate safety record.

Space has been provided to write in additional regulations.

TO THE STUDENT
How to Use This Student Workbook

This student workbook includes the material you need for keeping a record of your progess in carpentry.

Complete the following:

- Fill in the *Student Information* sheet in this workbook. This will help your teacher become better acquainted with you.
- Keep a record of your clean-up assignments in the space provided in this workbook.
- Make a list of the regulations that you are expected to follow.
- Read the *Safety Pledge* carefully. Sign it after you have agreed to obey the safety regulations.
- Use the *Accident Report* sheet to record needed information in case of an accident.
- Fill in the *Cost of Materials* chart. These costs will be used when making the bill of materials for each project.
- Make a plan for each project you build.
- Fill in the information at the top of each study guide. The textbook pages you should study are given at the top left side of the first page of each study guide. Write the answers in the blank spaces along the left side of each page.

You will find that this student workbook has six different kinds of questions. Samples of each kind of question are given on the next page. Study each type of question so you will know the correct way to answer each question.

Sample Questions

1. TRUE-FALSE (T-F). Read the statement carefully and decide if it is true or false. If the statement is true, put a T in the blank; if it is false, put an F.

   SAMPLE: Washington, D.C. is the capital of the United States. (T or F)

   T

2. MULTIPLE CHOICE WITH ONE RIGHT. Read the question and the possible answers. Select the best or correct answer and put the letter for this in the blank.

   SAMPLE: The following state is east of the Mississippi River *(one right)*: a. Arkansas; b. Missouri; c. Illinois; d. Iowa.

   C

3. MULTIPLE CHOICE WITH ONE WRONG. Study the question and the possible answers. Select the one that is wrong and put the letter for this in the blank.

   SAMPLE: The following are common fractions *(one wrong)*: a. 1/2; b. 3/8; c. 2; d. 3/4.

   C

4. COMPLETION. Study the sentence and decide what word will best finish the sentence. Write this in the space at the left.

   SAMPLE: The primary colors are red, yellow, and ____________.

   SAMPLE: The sun rises in the ____________ and sets in the west.

   SAMPLE: Milk is to cow as egg is to ____________.

   blue

   east

   chicken

5. MATCHING. Match each item in the left-hand column with the correct one in the right-hand column. Show your choice by placing the correct numbers in the spaces at the left.

   SAMPLE: Match the items at the left to the items at the right:

a. goal posts	1. football
b. hoop	2. tennis
c. puck	3. hockey
d. racquet	4. basketball

   a. 1
   b. 4
   c. 3
   d. 2

6. NAME OR IDENTIFICATION. Write in the space at the left the correct name of the item shown.

   SAMPLE: Name these common eating utensils:

   a

   b

   c

   a. knife
   b. fork
   c. spoon

STUDENT INFORMATION

Print:

1. Name ____
   Last First Middle

2. Home address ____

3. Home phone number ____

4. Year ____

5. School last attended ____

6. Parents' names:

   Father ____

   Mother ____

7. Parents' occupations:

   Father employed by ____

   Mother employed by ____

8. Hobbies or outside interests ____

   ____

9. Previous carpentry experience ____

10. Name of family doctor ____

    Phone ____

    Address ____

CLEAN-UP ASSIGNMENTS

	Date		Job
From ________________	to ________________		________________
From ________________	to ________________		________________
From ________________	to ________________		________________

SAFETY REGULATIONS

1. ____________________
2. ____________________
3. ____________________
4. ____________________
5. ____________________
6. ____________________
7. ____________________
8. ____________________
9. ____________________
10. ____________________

SAFETY PLEDGE

I pledge that I will follow all of the safety rules given in the book *Carpentry and Building Construction* and all of the regulations listed above. I will not use a power tool without first securing the permission of the instructor. I will report all accidents to the instructor immediately, no matter how small they are. I will help to maintain a safe environment by tending to my business and by not bothering other students who are busy.

Name ________________________________ Date ________________

ACCIDENT REPORT

1. Name of injured ____

   Address ____

   Telephone ____

2. Nature of injury (Cut, scratch, foreign matter in eye, etc.)

   ____

3. Tools or machines being used ________________________________

4. Witnesses to the accident: Name ____________________________

   Address ____________________________

   Name ____________________________

   Address ____________________________

5. Treatment: First aid __________________ By whom __________________

   Physician __________________ Address __________________

   Hospital __________________ Address __________________

6. Cause of accident (Poor condition, wrong procedure, etc.)

   ____

   ____

7. Correction (What will be done to prevent future accidents)

   ____

   ____

____

Name

COST OF MATERIALS

Lumber:

Kind	Cost Per Board Foot	Kind	Cost Per Board Foot
Pine			
Fir			
Maple			
Oak			
Birch			

Panel Stock:

Kind	Grade	Thickness	Cost Per Sq. Foot
Plywood			
Hardboard			
Particleboard			

COST OF MATERIALS (cont.)

Screws, Nails, and Other Items:

No. or Size	Kind	Cost Per _______
	Screws	
______________	_______	______________
______________	_______	______________
	Nails	
______________	_______	______________
______________	_______	______________
	Dowel	
______________	_______	______________
	Sandpaper	
______________	_______	______________
______________	_______	______________
	Hardware	
______________	_______	______________
______________	_______	______________
______________	_______	______________

Finishing Materials:

Kind	Cost Per _______

STUDY UNIT SCORES

STUDY UNIT NUMBER	POSSIBLE SCORE	NUMBER CORRECT	STUDY UNIT NUMBER	POSSIBLE SCORE	NUMBER CORRECT
1	42	______	31	35	______
2	46	______	32	25	______
3	38	______	33	34	______
4	59	______	34	44	______
5	52	______	35	34	______
6	42	______	36	41	______
7	47	______	37	57	______
8	169	______	38	24	______
9	32	______	39	26	______
10	60	______	40	30	______
11	41	______	41	38	______
12	29	______	42	62	______
13	35	______	43	38	______
14	61	______	44	21	______
15	51	______	45	30	______
16	28	______	46	21	______
17	32	______	47	42	______
18	53	______	48	64	______
19	43	______	49	32	______
20	29	______	50	38	______
21	25	______	51	41	______
22	33	______	52	42	______
23	33	______	53	46	______
24	42	______	54	27	______
25	43	______	55	65	______
26	28	______	56	49	______
27	39	______	57	23	______
28	43	______	58	21	______
29	33	______	59	33	______
30	23	______	60	28	______

To be used with Units 1 and 2 | Name ______________________

Pages 1-14 | Score: (42 possible) ______

Study Unit 1

Careers and Planning

______ 1. One of the largest industries in the United States is construction. (T or F)

______ 2. An example of light construction would be home building. (T or F)

______ 3. Some home building today is done with prefabricated components. (T or F)

______ 4. Three career levels that require special training and education are craft, ______, and professional.

______ 5. Examples of craft careers in building construction are *(one wrong)*: a. carpenter; b. architect; c. plumber; d. painter.

______ 6. There are fewer than one dozen skilled building trades. (T or F)

______ 7. A ______ is a craft worker in unions.

______ 8. According to Table 1-A, the growth in the number of manufacturing jobs is greater than the number of construction jobs. (T or F)

______ 9. There are over 800,000 carpenters employed in the construction trades. (T or F)

______ 10. The percentage change in construction from 1988-2000 will be 14.8%. (T or F)

______ 11. Specialization is common in carpenters' trades. (T or F)

______ 12. It is possible for a carpenter to learn the trade without formal education. (T or F)

______ 13. One who is entering construction by combining on-the-job training with basic instruction is called a ______.

______ 14. Courses like drafting, building techniques, and surveying are examples of preparation for ______ careers in building construction.

______ 15. A college degree is required for professional jobs in building construction. (T or F)

______ 16. Housing for a family should not cost more than ______ times the average annual income of the family.

17. Consider the following when choosing a lot for a house:

a. ______ a. Is it possible to use the ______ to supply part of heating and electrical needs?

b. ______ b. Are jobs, schools, community services and ______ facilities nearby?

c. ______ c. Is the lot shallow or ______?

d. ______ d. Is the neighborhood likely to remain relatively ______?

(Continued on next page)

_______________ 18. Not more than 25% of monthly income should be spent for all housing expenses, including mortgage payments, _______________, and repairs.

19. Three legal documents are needed before buying property:

a. _______________ a. The _______________ that shows the boundaries of the property.

b. _______________ b. The _______________ that is evidence of ownership.

c. _______________ c. The _______________ _______________ _______________ which is a history of deeds and other papers.

_______________ 20. House plans can be purchased from companies that specialize in designing stock house plans. (T or F)

_______________ 21. Purchasing stock plans is the least expensive way of obtaining house plans. (T or F)

_______________ 22. If the architect only designs the house, the fee is usually _______________ percent of the total building cost.

_______________ 23. A rough estimate of how much a house should cost can be found by multiplying the average cost of building per square foot in your locality by the number of square feet in the house plans. (T or F)

_______________ 24. To get an accurate price estimate of a home, ask a contractor to _______________ on the home.

_______________ 25. A mortgage loan is a _______________ on the property.

_______________ 26. Typical mortgages run from 12 to _______________ years.

_______________ 27. Charges for paperwork and similar items needed to make a loan are called _______________ costs.

_______________ 28. Parts of mortgage payment used to pay taxes and insurance are held in _______________.

_______________ 29. Contractors are usually paid a certain _______________ of the construction costs before the home is started.

_______________ 30. Building codes establish minimum standards of quality and _______________ in housing.

31. The three major building codes in the United States include the:

a. _______________ a. _______________ Building Code used in the West.

b. _______________ b. _______________ Building Code used in the Midwest.

c. _______________ c. _______________ Building Code used in the East.

_______________ 32. The electrical code used throughout the United States is the _______________ _______________ _______________.

_______________ 33. It is necessary to get a building _______________ before construction can begin.

_______________ 34. At key points during construction, a building _______________ will visit the job site to examine the work.

_______________ 35. If there are no problems after the house has been completed, a _______________ of Occupancy will be issued.

To be used with Unit 3
Pages 15-35

Name ______________________

Score: (46 possible) ______

Study Unit 2

Reading Prints

1. Designers and architects express their ideas in drawings by means of lines, symbols, and ________.
2. The ability to read and understand plans, drawings, and ________ is basic to all construction.
3. An exact copy of a drawing is called a ________.
4. The building industry makes use of prints called ________.
5. Blueprints do not fade. (T or F)
6. The metre is a metric symbol for ________.
7. A yard is a little shorter than a ________.
8. Metric units for length are *(one wrong)*: a. kilometre; b. millimetre; c. kilogram; d. centimetre.
9. One inch is equal to *(one right)*: a. 25.4 millimetres; b. 2.5 millimetres; c. 5.4 millimetres; d. 5.4 centimetres.
10. In building construction the inch is rounded off to ________ millimetres.
11. The litre is a measure for ________.
12. The litre is slightly larger than the ________.
13. The kilogram is a metric measure of ________.
14. A kilogram equals about *(one right)*: a. 1.8 pounds; b. 2.8 pounds; c. 2.2 pounds; d. 3.2 pounds.
15. In architectural drawings, metric measurements are given in ________.
16. Scale represents the ratio between the size of an object as drawn and its ________ size.
17. Scale is a unit of measurement. (T or F)
18. When a drawing is the same size as the object itself, it is called a ________ scale drawing.
19. A house plan that is drawn 3″ to 1′ is 1/4 full size. (T or F)
20. Figure 3-5 in the text shows two kinds of scales, the architects' scale and the ________ scale.
21. A perspective drawing, also called a ________ drawing, is a kind of pictorial drawing.
22. The kind of drawing prepared by the construction drafter is an architectural ________ drawing.

(Continued on next page)

23. Drawings are made of the following:

a. ______________, which show the shape.

b. Numbers telling sizes, called ______________.

c. Representations of things which would be difficult to draw, called ______________.

d. Written information and explanations, called ______________.

a. ______________ b. ______________ c. ______________ d. ______________

24. Identify the symbols for the kinds of materials given in Fig. 2-1.

a. ______________ b. ______________ c. ______________ d. ______________ e. ______________ f. ______________

a b c

d e f

Fig. 2-1.

______________ 25. Parts of a roof are *(one wrong)*: a. rafter; b. purlin; c. truss; d. girder.

______________ 26. Subflooring is laid directly over ______________.

27. Match the items on the left with their descriptions on the right:

a. specifications
b. sectional views
c. site plan
d. detail or detail drawings
e. building or floor plan
f. shop sketch
g. foundation or basement plan
h. elevations
i. framing plan
j. plot plan
k. bill of materials

1. table of information on needs for building a project
2. outline of lot
3. location of building on site
4. shows exact size of foundations and their location
5. cross-section view of house
6. show exterior view of house
7. simple drawing
8. shows structural members of house
9. written description of all materials
10. show the cross section of structural parts
11. show specific details of a house

a. ______________ b. ______________ c. ______________ d. ______________ e. ______________ f. ______________ g. ______________ h. ______________ i. ______________ j. ______________ k. ______________

______________ 28. All notes concerning a set of construction drawings should be placed on the drawings themselves. (T or F)

To be used with Unit 4
Pages 36-54

Name ____________________

Score: (38 possible) ______

Study Unit 3

Energy Use in the Home

______ 1. Energy consumed in living units is approximately one-______ of our total national energy consumption.

______ 2. The following are examples of exfiltration in the winter *(one wrong)*: a. leakage of warm air through cracks; b. leakage of warm air through windows; c. leakage of cold outside air into the house; d. leakage of warm air through doors.

______ 3. Heat is lost through water drains in the living unit. (T or F)

______ 4. Some heat is gained through solar energy, even in the standard house. (T or F)

______ 5. The heat that tends to flow into a structure is called ______.

______ 6. Nearly ______ percent of the heat loss in homes is due to exfiltration and infiltration.

______ 7. In cold weather, energy is needed to heat the air, clean it, and add ______ for comfort.

______ 8. Dark-colored roofs reduce solar heat gain. (T or F)

______ 9. The "payback" period for energy-saving devices is from one to ______ years.

______ 10. An all-purpose living area is called a great room. (T or F)

______ 11. Caulking is used to seal the house and prevent air leakage. (T or F)

______ 12. Most caulks and sealants come in a sealed ______ for use in a caulking gun.

______ 13. The temperature of the material to be caulked should be at least 60 degrees F. (T or F)

______ 14. The least expensive caulk is oil-based. (T or F)

______ 15. The approximate life span of silicone used outdoors is ______ or more years.

______ 16. Solvent acrylic is recommended for indoor use. (T or F)

______ 17. Polysulfide caulk resists chemicals. (T or F)

______ 18. All insulating sheathing is made from some type of plastic. (T or F)

______ 19. Sheet metal let-in braces are used with nonstructural insulating sheathing. (T or F)

______ 20. There are ______ basic kinds of insulating sheathing. (T or F)

______ 21. An insulating sheathing similar to the plastic used to make coffee cups is ______.

(Continued on next page)

_______________ 22. An insulating panel that is faced on both sides with foil is _______________.

_______________ 23. Energy efficiency in a standard house can be improved by *(one wrong)*: a. using the correct kind of sheathing; b. adding the proper insulation; c. increasing the square footing of the house; d. using the proper kind of storm windows and doors.

_______________ 24. Attic vents may be in the roof, _______________, or gable.

_______________ 25. The R-2000 home program is sponsored by the _______________ Home Builders Association.

_______________ 26. The energy-saving home sponsored by the U.S. Department of Housing and Urban Development is known as the _______________ plan.

_______________ 27. Homes in hot, humid climates should have windows that face east and west. (T or F)

_______________ 28. Homes in hot, humid climates usually have a slab-on-grade foundation. (T or F)

_______________ 29. Homes in hot, humid climates should use more framing members to reduce heat conduction. (T or F)

_______________ 30. The Arkansas plan house has _______________ inches of insulation in the walls and _______________ inches in the ceiling.

____

_______________ 31. In the Arkansas plan house, about _______________ percent of the energy saving was achieved by using the correct windows and doors.

_______________ 32. The two kinds of solar energy systems for living units are passive and _______________.

_______________ 33. In an active solar energy system, the major problems are posed by the _______________ and storage units.

_______________ 34. In the passive system, heat circulates to the living space by conduction, _______________ and radiation.

_______________ 35. There are _______________ basic elements in a complete passive system.

_______________ 36. There are three types of passive solar energy systems – direct gain, indirect gain, and _______________ gain.

_______________ 37. With a photovoltaic panel, electricity can be produced directly from _______________.

To be used with Unit 5

Pages 56-73

Name ______________________

Score: (59 possible) ______

Study Unit 4

Wood as a Building Material

______________ 1. The substance located under the bark of trees that produces new wood is the ______________.

______________ 2. The sap is carried to leaves by the ______________.

______________ 3. A tree's age can be judged by counting the number of growth ______________.

______________ 4. Softwoods, when classified by species, are sometimes harder than hardwoods. (T or F)

______________ 5. Another name for hardwoods is ______________ trees.

______________ 6. Softwoods come from trees that shed their leaves. (T or F)

______________ 7. Earlywood is darker than latewood. (T or F)

______________ 8. Common hardwoods include *(one wrong)*: a. fir; b. maple; c. cherry; d. oak.

______________ 9. Plain-sawed lumber does not tend to vary as much as quarter sawed. (T or F)

______________ 10. One common method for cutting boards is called plain sawed for hardwood and ______________ grained for softwood.

______________ 11. The two common methods for drying lumber are ______________ drying and air drying.

______________ 12. When seasoning rough lumber, stickers should be placed between the pieces. (T or F)

______________ 13. Oven drying of lumber takes longer than the natural method. (T or F)

______________ 14. After lumber is dried in air, the moisture content should be ______________ percent or less.

______________ 15. Lumber dried in an oven should have a moisture content of less than ______________ percent.

______________ 16. Lumber shrinks about the same in length as it does in width. (T or F)

______________ 17. When wood contains enough water to saturate the cell walls it is said to be at the fiber ______________ point.

______________ 18. Once wood has been dried, its moisture content remains constant. (T or F)

______________ 19. Moisture content of wood can be determined by the oven drying method. (T or F)

______________ 20. Another way to check moisture content is with a moisture ______________.

(Continued on next page)

______ 21. When a board is dried rapidly in an oven, the center will have *(one right)*: a. more moisture than the outside of the board; b. less moisture than the outside; c. the same amount of moisture; d. no moisture.

______ 22. Pieces of higher and lower quality are found in the same grade of lumber. (T or F)

______ 23. The best-quality lumber is found in the ______ part of the tree trunk, near the outside.

24. Match the items on the left with the descriptions on the right:

a. ______
b. ______
c. ______
d. ______
e. ______
f. ______
g. ______
h. ______
i. ______
j. ______
k. ______
l. ______
m. ______
n. ______
o. ______

a. stain
b. knot
c. check
d. pitch-pocket
e. decay
f. wane
g. cup
h. crook
i. shake
j. warp
k. summerwood
l. split
m. pitch
n. torn grain
o. bow

1. any variation from true surface
2. deviation flatwise from straight line
3. discoloration
4. dense, outside part of ring
5. branch or limb embedded in tree
6. caused by fungi
7. grain separation between or through growth rings
8. bark on edge or corner
9. lengthwise grain separation due to seasoning
10. separation of wood completely through to opposite surface
11. edgewise deviation that goes full length
12. an opening containing pitch or bark
13. deviation flatwise across width
14. wood torn out
15. accumulation of resin

______ 25. There are ______ grades of hardwood.

______ 26. When a tree is first cut down, it can contain moisture ranging from 30 to ______ percent more than it does after drying.

______ 27. All wood should be dried the same amount regardless of its use. (T or F)

______ 28. Lumber labeled according to its strength and loadbearing quality is called ______ grade lumber.

29. Match the items at the left with the descriptions on the right:

a. ______
b. ______
c. ______
d. ______
e. ______
f. ______
g. ______
h. ______

a. tensile stress
b. compression
c. live load
d. impact load
e. elasticity
f. static load
g. shear stress
h. fatigue load

1. applies to furniture in a home
2. tends to make a piece longer
3. squeezing or crushing
4. applied a large number of times
5. forces that meet head on
6. a sudden, sharp force
7. weight applied to a house
8. stiffness of a piece and its resistance to bending

______________________ 30. Painting wood prevents it from taking on moisture. (T or F)

______________________ 31. If lumber is stored properly, it will not absorb much moisture. (T or F)

______________________ 32. Hardwood is sold in standard widths and lengths. (T or F)

______________________ 33. The nominal size of lumber and its actual size are the same. (T or F)

______________________ 34. A board that measures 2″ × 6″ × 12′ contains __________ board feet.

______________________ 35. A 2″ × 4″ that is 12′ long contains __________ board feet.

______________________ 36. In the metric system, the unit for thickness and width of lumber is the __________.

______________________ 37. The metric unit for length is the __________.

______________________ 38. The metric measurement for a standard plywood sheet is 1220 mm by __________ mm.

To be used with Units 6 and 7 Name ______________________

Pages 74-90 Score: (52 possible) ______

Study Unit 5

Plywood and Composition Panels

______ 1. A thin sheet of wood that is sliced, sawed, or peeled from a log is called ______.

______ 2. Plywood is made from layers and/or ______ of veneer or veneer and wood.

______ 3. The plywood layers are glued together with the grain running in the same direction. (T or F)

______ 4. There are two major types of hardwood plywood, namely, veneer core and ______ core.

______ 5. Logs or flitches must be softened or tenderized before cutting the veneer. (T or F)

______ 6. Most plywood used in the construction industry is made from softwood. (T or F)

______ 7. The three kinds of plywood construction are veneer core, lumber core, and ______ core.

______ 8. Exterior plywood must be put together with ______ glue.

______ 9. Exterior plywood construction requires the application of heat. (T or F)

______ 10. Interior plywood can be made without heat. (T or F)

______ 11. Plywood that is made for special engineering use and meets such properties as tension and compression is ______ plywood.

______ 12. Plywood is always manufactured in an odd number of layers. (T or F)

______ 13. Plywood grades range from N to ______.

______ 14. The 70 different species of wood from which plywood is made are divided into ______ groups.

______ 15. Maple, birch, and Douglas fir are three of the strongest woods. (T or F)

______ 16. Choose plywood thickness according to its use. (T or F)

______ 17. The number of plies and the thickness of plywood are important factors in its use. (T or F)

______ 18. The front and back of a plywood piece can have different grades. (T or F)

______ 19. If plywood is stored on edge, it will ______.

______ 20. It is best to store plywood flat. (T or F)

______ 21. To saw plywood by hand, place the good face down. (T or F)

______ 22. Screws and nails hold well only in the face of plywood. (T or F)

(Continued on next page)

_______________ 23. The correct nail and screw size to use depends on the thickness of the plywood. (T or F)

_______________ 24. Composition panels are made from pieces of wood. (T or F)

_______________ 25. Materials for composition panels can be made from trees that are too small for lumber. (T or F)

_______________ 26. Urea resins are waterproof. (T or F)

_______________ 27. OSB is used primarily for siding and subflooring. (T or F)

_______________ 28. OSB will not shrink or swell with a change in humidity. (T or F)

_______________ 29. Waferboard is made of flakes of wood that are randomly aligned throughout the panel. (T or F)

_______________ 30. Hardboard is an all-wood panel manufactured from wood fibers. (T or F)

_______________ 31. Some characteristics of hardboard are *(one wrong)*: a. exceptional strength; b. superior wear resistance; c. hard to work with ordinary tools; d. easy to paint.

_______________ 32. The three types of hardboard are standard, _______________, and service.

_______________ 33. Hardboard classified as S1S has one surface smooth and the other rough. (T or F)

_______________ 34. Manufactured boards have the following advantages over basic building materials like wood *(one wrong)*: a. they can be made to more sizes, shapes, and surfaces; b. they are heavier; c. they are high in insulating values; d. they do not tend to split, crack, or splinter.

_______________ 35. Hardboard, particleboard, plywood, insulating board, gypsum board, and plastic foam board are all examples of building boards. (T or F)

36. Match the hardboards at the left with the descriptions at the right:

a. _______________ a. embossed
b. _______________ b. wood-grain
c. _______________ c. perforated
d. _______________ d. underlayment
e. _______________ e. exterior siding
f. _______________ f. acoustical
g. _______________ g. filigree

1. most popular for indoor paneling
2. made to imitate a different material
3. for sound control
4. excellent for storage or display
5. cut into patterns
6. a good subflooring
7. prefinished and primed at the factory

_______________ 37. Particleboard is made by combining wood flakes with adhesives. (T or F)

_______________ 38. Heat is not required in the manufacture of particleboard. (T or F)

_______________ 39. The two kinds of particleboard are mat-formed and _______________.

_______________ 40. Particleboard has the following characteristics *(one wrong)*: a. is a good base for plastic laminates; b. does not warp; c. takes a good paint finish; d. is stiff and strong.

_______________ 41. Particleboard is used in home construction to make kitchen counters, sink tops, and cabinets. (T or F)

______________________ 42. The largest single use of particleboard in the construction industry is for floor ____________.

______________________ 43. There is a wide variation in the types of particleboard manufactured. (T or F)

______________________ 44. Particleboard can be worked with standard wood tools. (T or F)

______________________ 45. Particleboard is always purchased prefinished. (T or F)

______________________ 46. Most ordinary wood joints can be made in particleboard. (T or F)

To be used with Unit 8 Name ______________________

Pages 91-100 Score: (42 possible) ______

Study Unit 6

Framing Connectors and Engineered Wood

______ 1. Several important structural products include *(one wrong)*: a. laminated-veneer lumber; b. heavy timbers; c. glue-laminated beams; d. metal connectors.

______ 2. Metal joist hangers can replace toenailed connections. (T or F)

______ 3. A family of products made with wood veneer as the basic element is called ______ ______ lumber.

______ 4. The letters used to identify laminated-veneer lumber are ______.

______ 5. LVL products are available in lengths of ______ feet or more.

______ 6. LVL products shrink and swell less than solid lumber. (T or F)

______ 7. LVL products are cross-laminated. (T or F)

______ 8. An I-beam is also called an ______.

______ 9. I-beams can be used only in floor construction. (T or F)

______ 10. The webs of LVL I-beams are made from plywood or oriented-strand board. (T or F)

______ 11. LVL I-beams use contact adhesive to attach webs and chords. (T or F)

______ 12. The easiest way of cutting an I-beam is with a ______ ______ saw.

______ 13. I-beams used in floor construction can be nailed to the plate. (T or F)

______ 14. I-beams can be secured with metal ______ ______.

______ 15. Nails should be driven sideways (parallel to the laminations) into an I-beam chord. (T or F)

______ 16. I-beams can be tripled to form a header. (T or F)

______ 17. The web of a wood I-beam has ______ inch diameter pre-scored knockouts located approximately 12″ on center. (T or F)

______ 18. Knockouts in I-beams are used to create passages for ______ and ______ lines.

______

______ 19. Framing connectors of 12-gauge galvanized steel are approximately ______ inch (fractional) thick.

______ 20. The chords of I-beams can be notched or drilled as needed. (T or F)

______ 21. LVL headers are ______ in cross-section.

______ 22. Holes can be cut in LVL headers. (T or F)

______ 23. All laminated-veneer products use ______ adhesive.

(Continued on next page)

______ 24. Glue-laminated beams are also called ______.

______ 25. Glue-laminated beams are resistant to fire. (T or F)

______ 26. Glue-laminated beams are made of layers of solid sawn lumber glued together. (T or F)

______ 27. Glue-laminated beams can be used for *(one wrong)*: a. garage door; b. windows; c. patio door; d. stair stringer.

______ 28. A slight upward curve in the glulam, like the crown in a piece of lumber is called ______.

______ 29. Three grades of glulam include *(one wrong)*: a. industrial; b. commercial; c. architectural; d. premium.

______ 30. Premium-grade glulams are used where appearance is of prime importance. (T or F)

______ 31. Glulams used for standard headers are either ______ inches or ______ inches wide.

______

______ 32. A metal connector is a formed or stamped metal bracket. (T or F)

______ 33. The builder must check local building codes to make sure certain metal connectors are approved for that application. (T or F)

______ 34. A joint hanger is the most common metal connector. (T or F)

______ 35. Use ______ nails to improve the strength of joist hangers.

______ 36. Figure 6-1 shows a metal connector between the ______ and a double top plate.

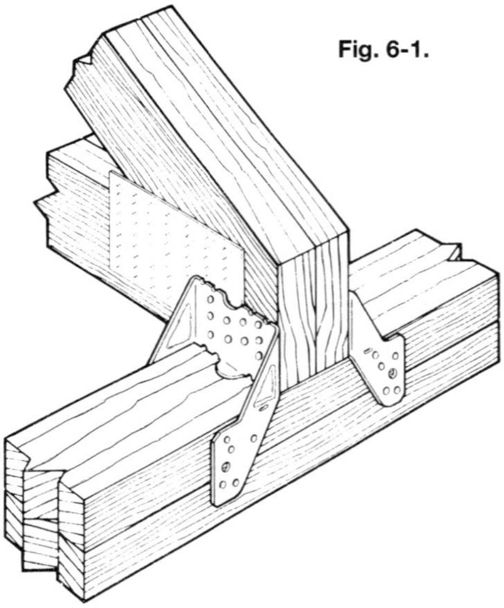

Fig. 6-1.

______ 37. There are two methods of installing a joist hanger. (T or F)

______ 38. The most common mistake when installing joist hangers is to use too many nails. (T or F)

______ 39. Two common shapes of metal framing ties are the flat strap and the ______ shape.

______ 40. Drywall screws are recommended to secure metal framing connectors. (T or F)

To be used with Unit 9

Pages 102-108

Name ____________________

Score: (47 possible) ______

Study Unit 7
Safety

______ 1. Injuries are costly because they reduce the efficiency of the work force. (T or F)

______ 2. Falling is the most common cause of injury on the job site. (T or F)

______ 3. Older workers are more likely to be injured than younger workers. (T or F)

______ 4. Follow this rule: "Lift with your back, not with your knees." (T or F)

______ 5. Many building materials become slippery when wet or frosty. (T or F)

______ 6. To protect yourself against chemical fumes, wear a ______.

______ 7. Tilesetters should wear kneepads. (T or F)

______ 8. When stock being cut on a table saw is thrown back at a high speed, this is called ______.

______ 9. When using a portable circular saw, force the saw into the cut. (T or F)

______ 10. Moisture can turn many materials into good conductors. (T or F)

______ 11. GFCI is a fast-acting circuit breaker. (T or F)

______ 12. Proper eye protection requires wearing *(one wrong)*: a. a mask; b. safety goggles; c. sunglasses; d. a safety shield.

______ 13. The correct dress for building construction includes *(one wrong)*: a. jewelry; b. sleeves of shirts or jackets buttoned; c. short hair or a hair protector; d. working attire.

______ 14. Complete safety attire for building construction includes the following *(one wrong)*: a. hard hat; b. safety shoes; c. safety eye protector; d. tie.

______ 15. When lifting heavy objects, use the muscles in your back. (T or F)

16. Follow these general safety practices:

a. ______ a. Always walk; do not ______.

b. ______ b. Never talk to or ______ anyone who is working on a machine.

c. ______ c. Remove the power plug or turn off the ______ supply to a machine when changing cutters or blades.

d. ______ d. Never leave ______ or pieces of stock lying on the table surface of a machine before using them.

e. ______ e. When finished with a machine, turn off power and wait until the cutter has come to a complete ______ before leaving.

f. ______ f. Always carefully check stock for knots, splits, ______ objects and other defects before machining.

(Continued on next page)

g. ______________________

h. ______________________

i. ______________________

j. ______________________

k. ______________________

l. ______________________

m. ______________________

n. ______________________

o. ______________________

g. Do not use a machine until you ____________ it thoroughly.

h. Use ____________ on power equipment.

i. Always keep your ____________ away from the moving cutting edges.

j. Keep the floor around the machine ____________.

k. Make all adjustments with the power ____________.

l. Always use a ____________ to clean the table surface.

m. Keep your eyes focused on where the ____________ action is taking place.

n. Make sure that tools are ____________.

o. Report strange ____________ or faulty operation of machines to the instructor.

17. Complete the following about avoiding falls:

a. ______________________

b. ______________________

c. ______________________

a. Learn to watch your ____________; avoid objects that could trip you.

b. Check ____________ and temporary walkways before walking on them.

c. Use only ____________ which are in good condition and set up properly.

______________________ 18. The correct abbreviation for the Federal Occupational Safety and Health Act is ____________.

______________________ 19. The Federal Occupational Safety and Health Act was passed by the Congress of the United States in April of ____________.

______________________ 20. Scraps and rubbish at the job site should be removed weekly. (T or F)

______________________ 21. Materials and equipment should be kept in neat, straight stacks. (T or F)

______________________ 22. A board with protruding nails should have the nails bend downward. (T or F)

______________________ 23. Aisles and walkways should be clear of tools, materials, and debris. (T or F)

______________________ 24. Long pieces of material should be carried by *(one right)*: a. one person; b. two people; c. three people; d. several people.

______________________ 25. The recommended extension cord size for use with portable power tools when the tool has an amperage rating of 8 and the cord length is 100 is ____________ feet.

a. ______________________________
b. ______________________________
c. ______________________________
d. ______________________________
e. ______________________________

26. Complete the following concerning safety rules for portable power tools:
    a. Never use portable power tools in contact with ____________.
    b. Portable power tools should be properly ____________.
    c. Always wear approved ____________ protection.
    d. Always disconnect the power ____________ when the work is completed.
    e. Be sure that the ____________ is in the off position before connecting to a power plug.

____

27. Whenever the drawing shown here in Fig. 7-1 appears with an illustration, a ____________ must be used for the operation shown.

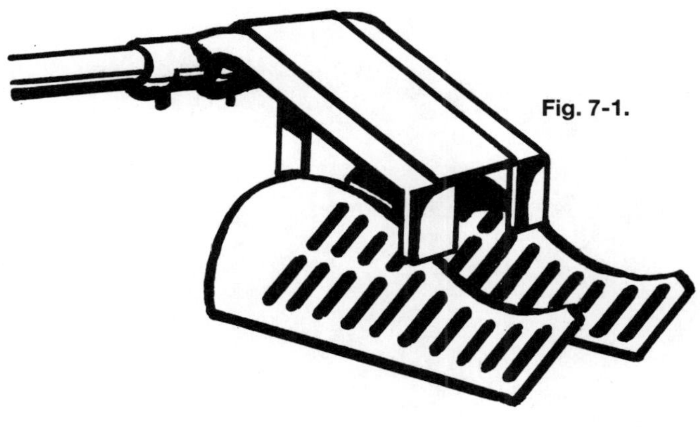
Fig. 7-1.

(Continued on next page)

Name ______________________________

Score: (169 possible) ______

Study Unit 8

Hand Tools

LAYOUT, MEASURING, AND CHECKING DEVICES

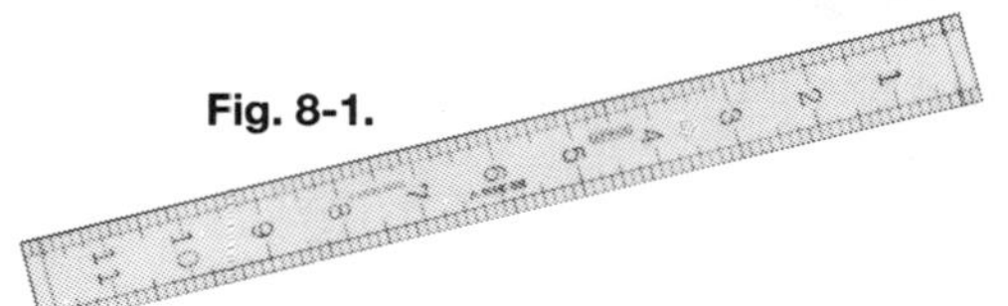

Fig. 8-1.

1. The hand tools shown here are numbered 8-1 through 8-6. Place these numbers in the blanks provided, matching tools with their correct names, descriptions, and uses.

_________ Zig-zag rule

_________ Bench rule

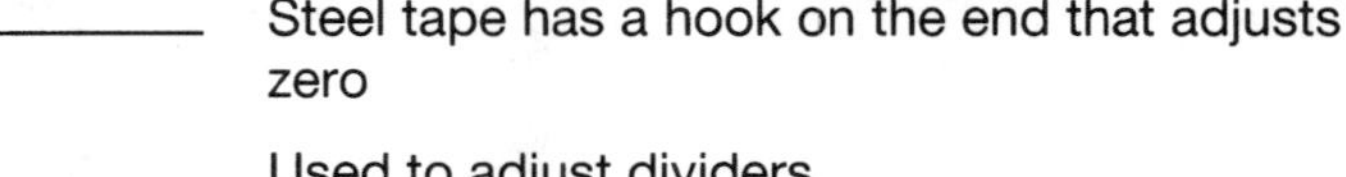

_________ Steel tape has a hook on the end that adjusts to true zero

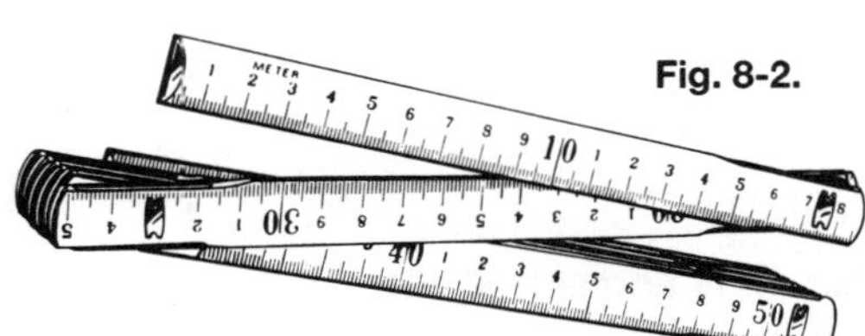

Fig. 8-2.

_________ Used to adjust dividers

_________ Try square

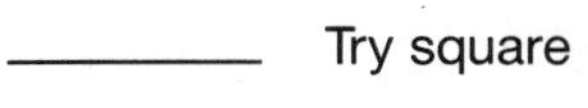

_________ A folding rule

_________ A good tool for measuring an angle

_________ Sliding T bevel

_________ This should never be used as a straightedge

_________ Combination square

_________ Used to measure or transfer an angle between 0 and 180 degrees

Fig. 8-3.

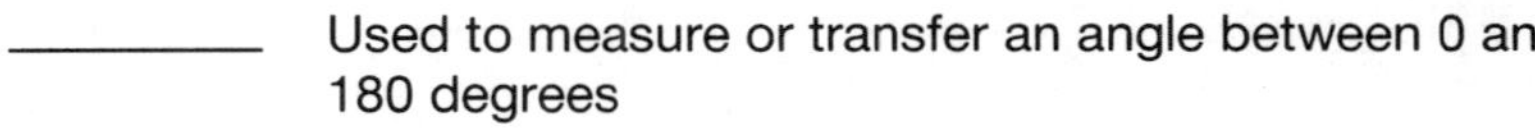

_________ Flexible tape rule

_________ The blade on this tool slides along the handle

_________ Comes in lengths from 6′ to 100′ (2 m to 50 m)

_________ Used to mark and test a 45-degree miter

_________ Measures irregular and regular shapes

_________ Used to check adjacent surfaces for squareness

_________ Measures distances, using tool flat or on edge

_________ Makes simple measurements

_________ Makes accurate inside measurements

_________ Makes lines across face or edge of stock

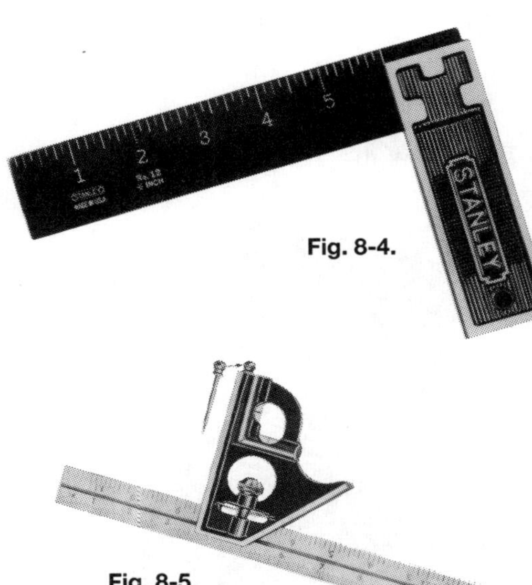

Fig. 8-4.

Fig. 8-5.

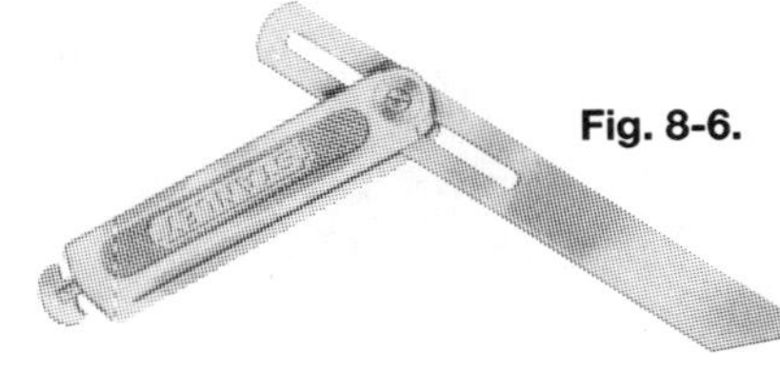

Fig. 8-6.

(Continued on next page)

2. The hand tools shown here are numbered 8-7 through 8-14. Place these numbers in the blanks provided, matching tools with their correct names, descriptions, and uses.

_________ Scratch awl

_________ Determines the corners of buildings

_________ Used to step off measurements

_________ Used to check squareness

_________ Framing or rafter square

_________ Checks levelness and plumbness

_________ Dividers

_________ A pointed tool with a handle

_________ Used to lay out rafters and stairs

_________ Marks parallel lines

_________ Trammel points

_________ Large steel square

_________ Used to locate points and scribe lines

_________ Carpenter's level

_________ Scribes arcs and circles larger than those made with dividers

_________ Establishes vertical lines

_________ Marking gauge

_________ Plumb bob and line

_________ Two-legged tool for laying out a circle

_________ Tool with a head, point, and beam

_________ Metal leg on tool can be replaced with pencil

_________ Metal pointers can be fastened to a long bar of wood or metal

_________ A metal weight with a pointed end

_________ Calculates volumes and areas

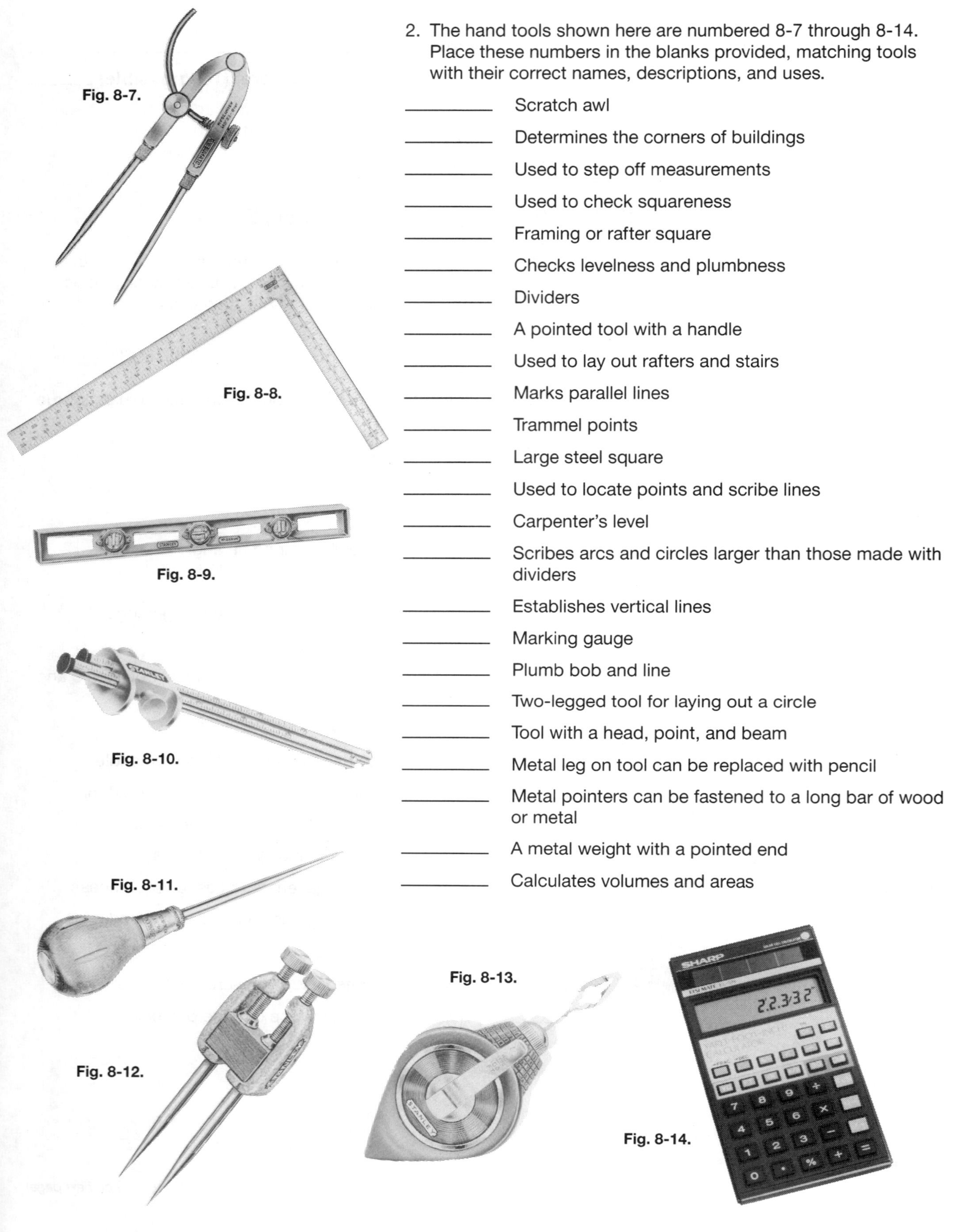

Fig. 8-7.

Fig. 8-8.

Fig. 8-9.

Fig. 8-10.

Fig. 8-11.

Fig. 8-12.

Fig. 8-13.

Fig. 8-14.

SAWING TOOLS

3. The saws shown here are numbered 8-15 through 8-22. Place these numbers in the blanks provided, matching the saws with their correct names, descriptions, and uses.

________ Rip saw

________ Use this saw to cut with the grain

________ Cuts gentle and inside curves

________ Back saw

________ A good saw for cutting with or across the grain

________ Coping saw

________ Can be used in a miter box, although shorter than a miter box saw

________ Has an extremely thin blade with very fine teeth

________ Compass saw

________ The best saw for tiny openings

________ A 10″ or 12″ narrow taper saw

________ Dovetail saw

________ Its tapered blade helps in cutting a circle

________ Crosscut saw

________ A U-shaped saw for scroll work

________ Miter box saw

________ Excellent for fine joint cutting

________ Keyhole saw

________ A long back saw

Fig. 8-15.

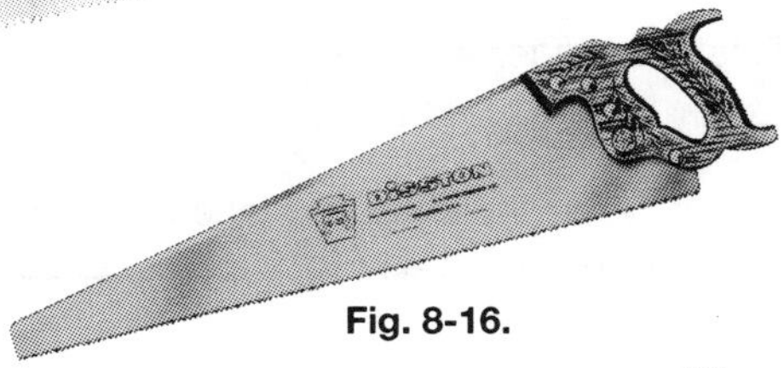

Fig. 8-16.

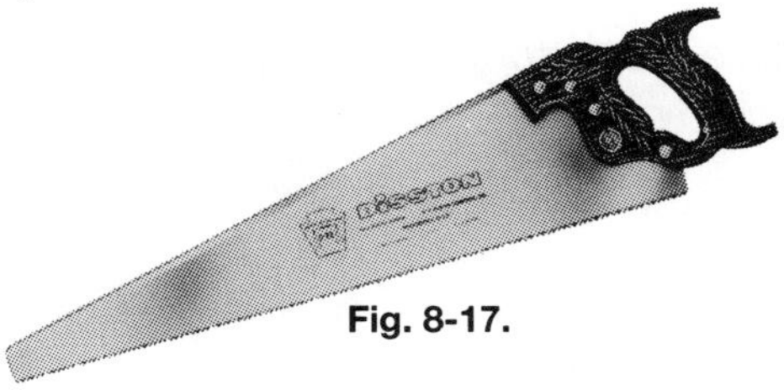

Fig. 8-17.

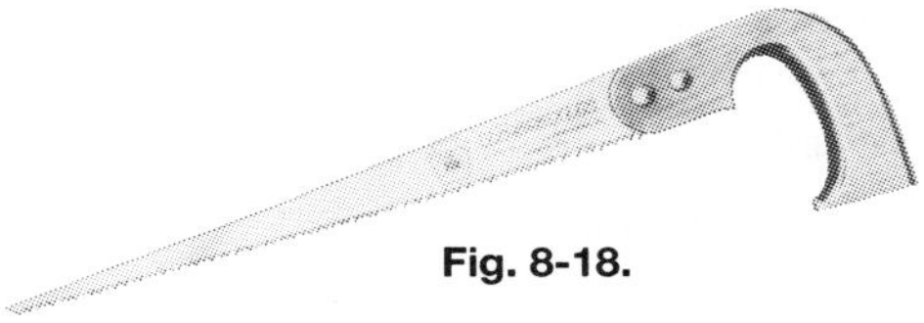

Fig. 8-18.

Fig. 8-19.

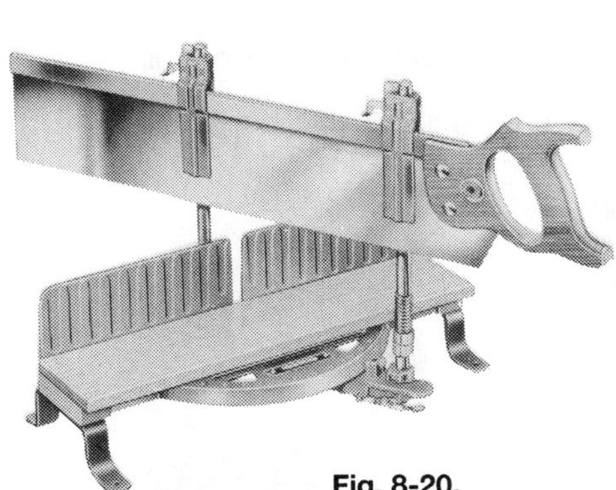

Fig. 8-20.

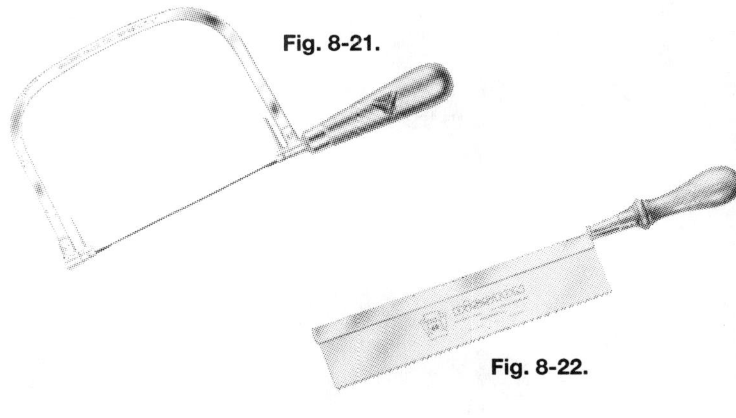

Fig. 8-21.

Fig. 8-22.

(Continued on next page)

EDGE-CUTTING TOOLS

Fig. 8-23.

4. The tools shown here are numbered 8-23 through 8-32. Place these numbers in the blanks provided, matching the tools with their correct names, descriptions, and uses.

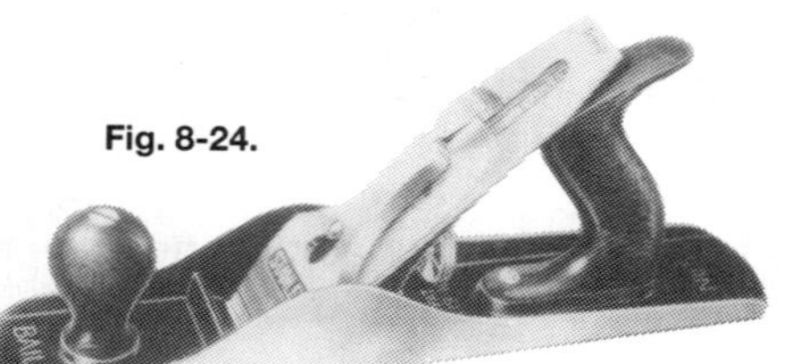

Fig. 8-24.

Fig. 8-25.

Fig. 8-26.

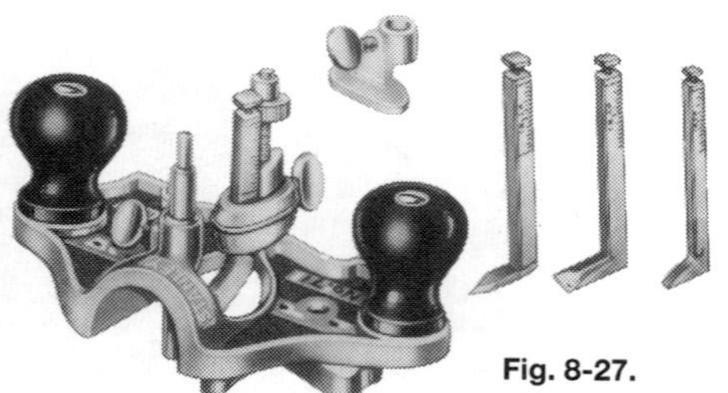

Fig. 8-27.

_________ Fore plane

_________ Trims and shapes wood

_________ Block plane

_________ For fine flat finish or longer surfaces

_________ Used to cut and trim wood

_________ To hold a piece of wood, you would place it in a vise, which is shown in Question 5 as Fig. _____________

_________ For smaller work

_________ Used to surface the bottom of grooves

_________ Chisel

_________ An 18″ plane

_________ Router plane

_________ Good for smoothing the edge of a door

_________ Hatchet

_________ Used to trim pieces

_________ To smooth and flatten edges

_________ Used to cut and trim veneer

_________ Used to surface bottom of dadoes

_________ An all-purpose knife used to make accurate layouts

_________ For planing the ends of molding

_________ Shown in Question 5, this could be used to hold wood for planing.

_________ Ideal for rough surfaces

_________ Jack plane

_________ Cuts grooves and shapes irregular openings

_________ A 7″ to 9″ plane for general use

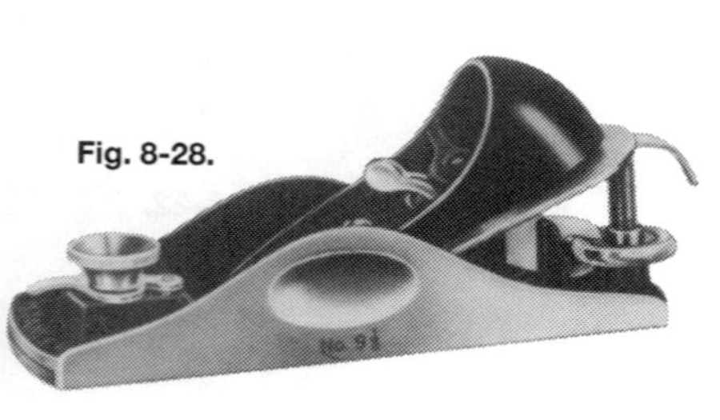

Fig. 8-28.

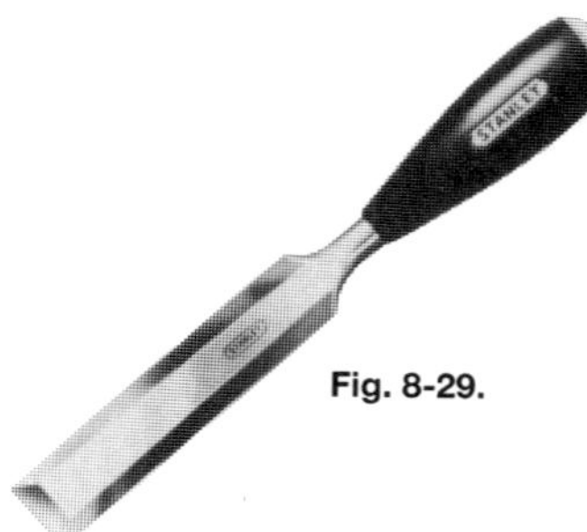

Fig. 8-29.

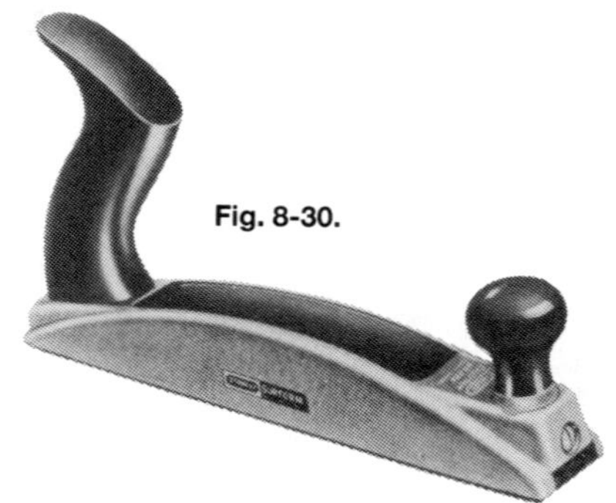

Fig. 8-30.

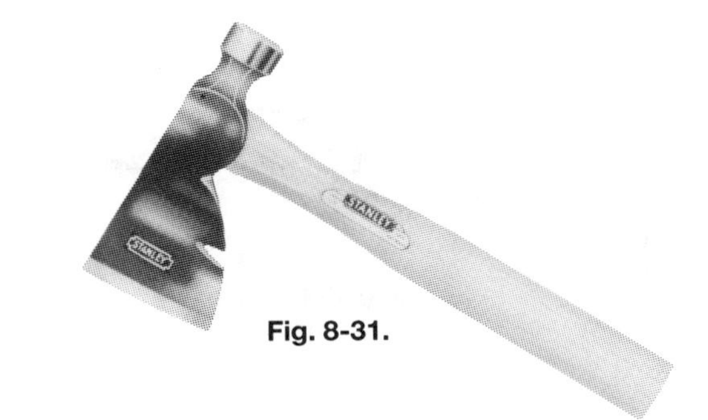

Fig. 8-31.

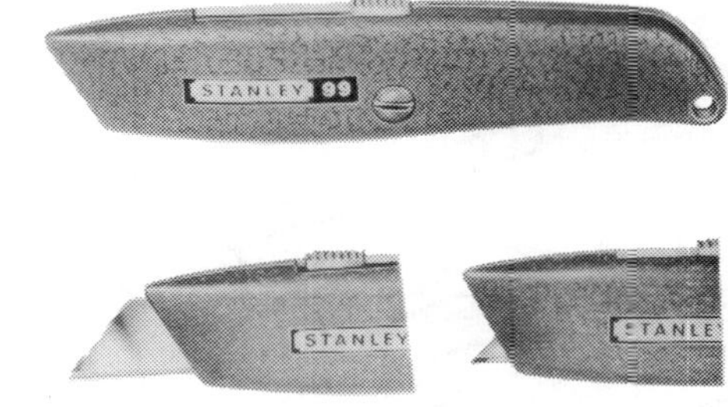

Fig. 8-32.

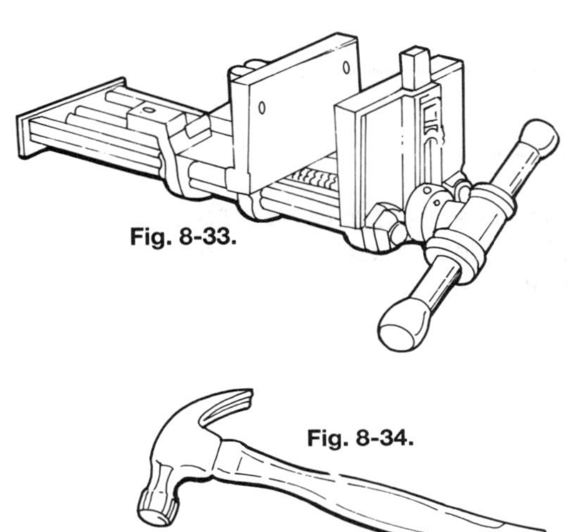

Fig. 8-33.

5. The tools shown here are numbered 8-33 through 8-41. Place these numbers in the blanks provided, matching the tools with their correct names, descriptions, and uses.

________ For driving nails below wood surfaces

________ Screwdriver

________ Has claw designed primarily for removing nails

________ Vise

________ Striking tool used when steel face would damage the surface of the material

________ Comes with slotted or Phillips head

________ Pneumatic nailer-stapler

________ Ripping bar

________ Hammer designed to pry nailed pieces apart

________ May attach permanently to workbench

________ Nail set

________ Mallet

________ Drives staples with spring-driven plunger

________ Claw hammer

________ Rip hammer

________ For attaching insulation

________ Available in lengths up to 8′

________ Holds work for planing

Fig. 8-34.

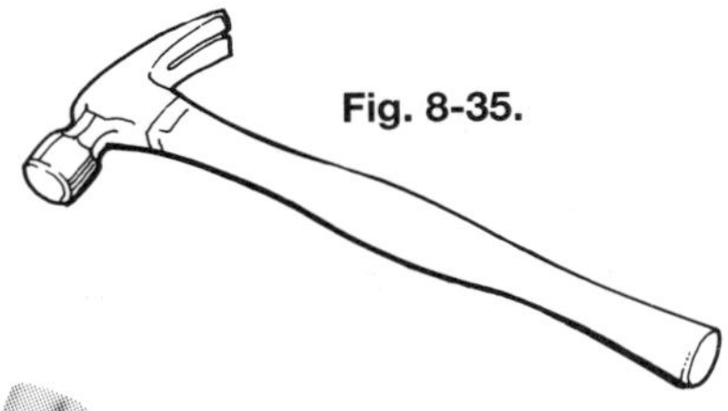

Fig. 8-35.

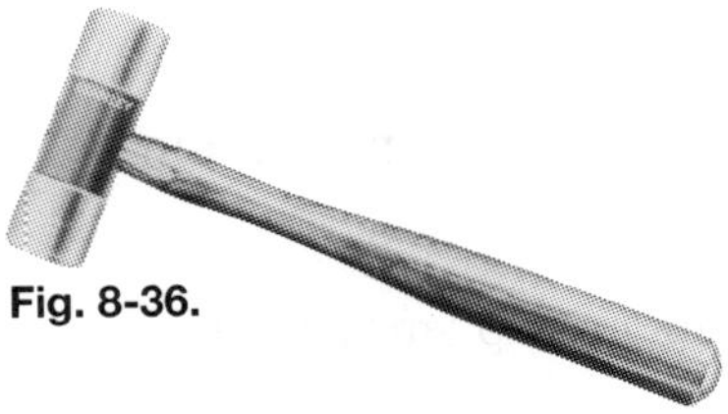

Fig. 8-36.

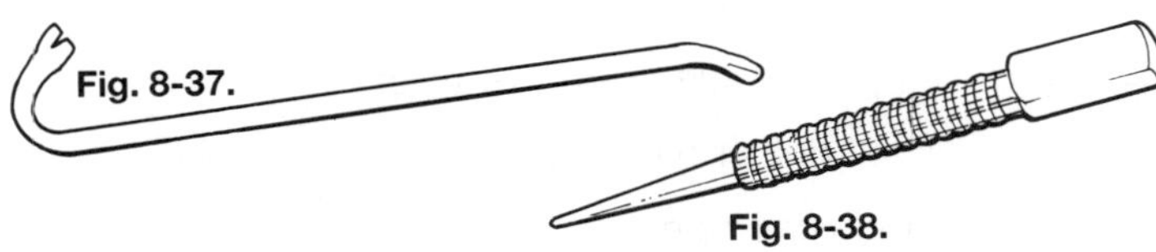

Fig. 8-37.

Fig. 8-38.

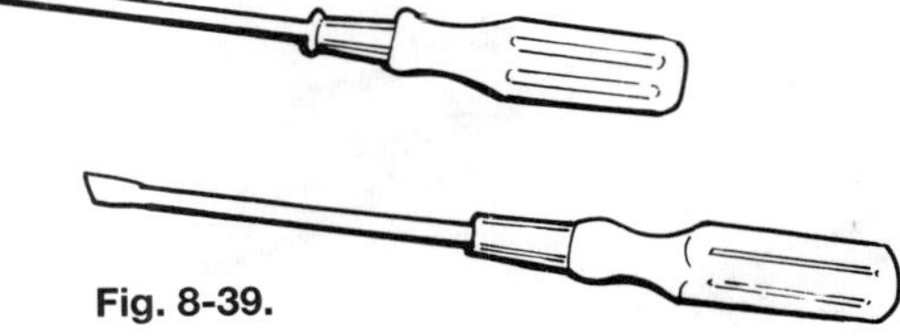

Fig. 8-39.

(Continued on next page)

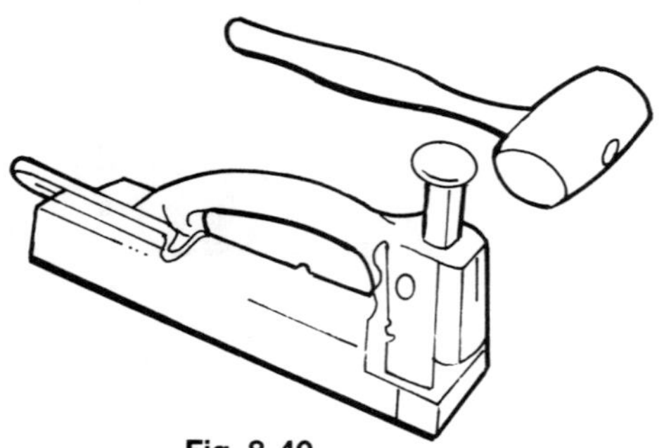
Fig. 8-40.

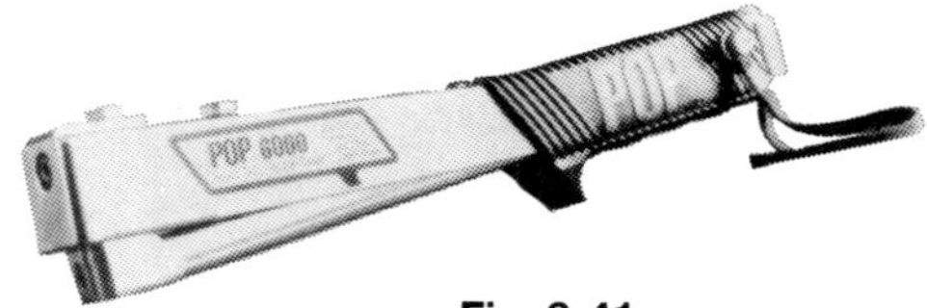

Fig. 8-41.

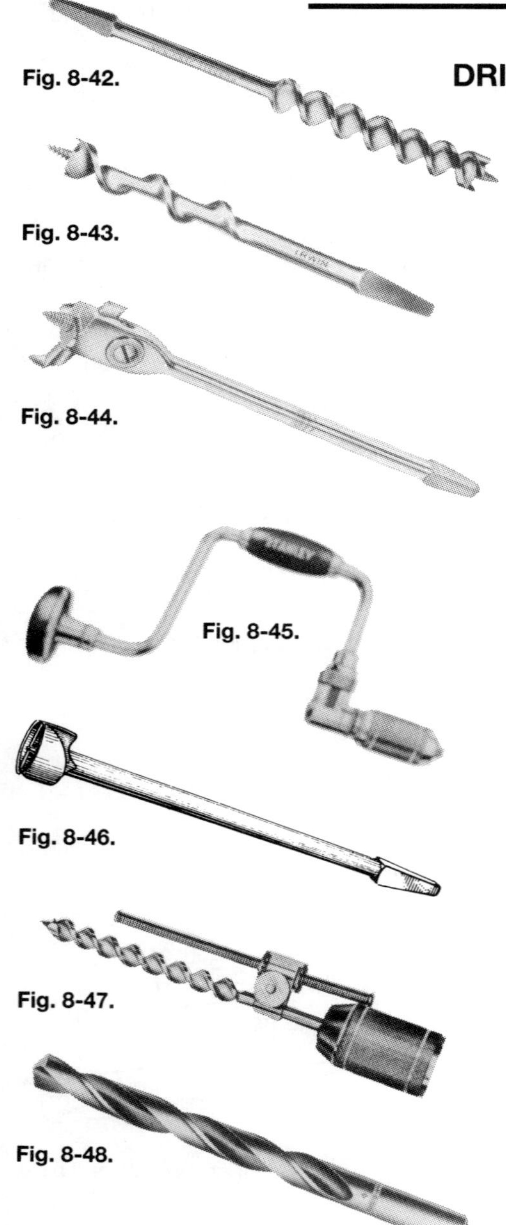
Fig. 8-42.
Fig. 8-43.
Fig. 8-44.
Fig. 8-45.
Fig. 8-46.
Fig. 8-47.
Fig. 8-48.

DRILLING AND BORING TOOLS

6. The tools shown are numbered 8-42 through 8-50. Place these numbers in the blanks provided, matching the tools with their correct names, descriptions, and uses.

_________ Foerstner bit

_________ Bores holes for making dowel joints

_________ Has a 3-jaw chuck

_________ Auger bit

_________ Bores holes larger than 1″

_________ Plain and ratchet types

_________ Hand drill

_________ Bores holes 1/4″ or larger

_________ Solid clamp and spring types

_________ A tool with drill points and handle for making many small holes

_________ Brace

_________ Will bore a shallow hole with a flat bottom

_________ Bit or depth gauge

_________ Holds twist-drills for drilling small holes

_________ Automatic drill

_________ Expansion bit

_________ May be single or double twist

_________ Twist drill or bit stock drill

_________ Holds and operates bits

_________ Dowel bit

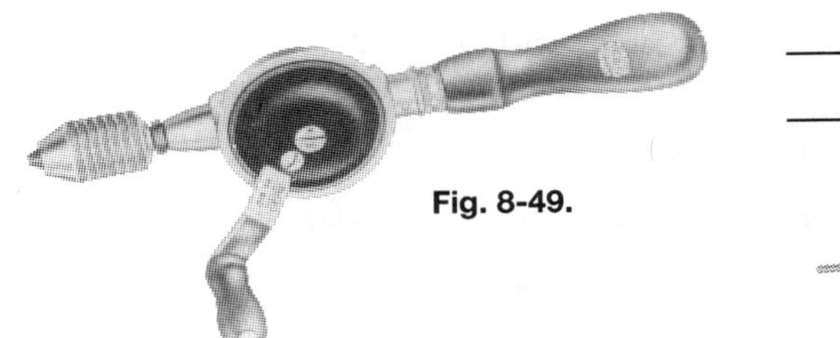

Fig. 8-49.

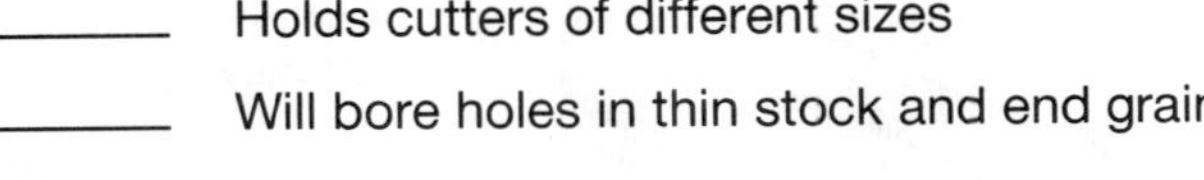

______ Holds cutters of different sizes

______ Will bore holes in thin stock and end grain

Fig. 8-50.

METALWORKING TOOLS

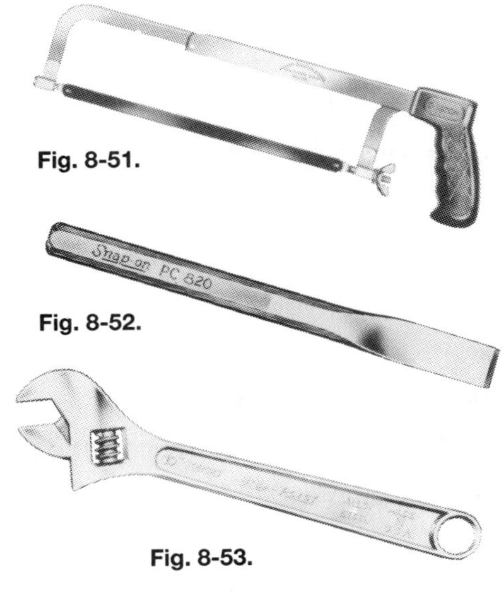

Fig. 8-51.

Fig. 8-52.

Fig. 8-53.

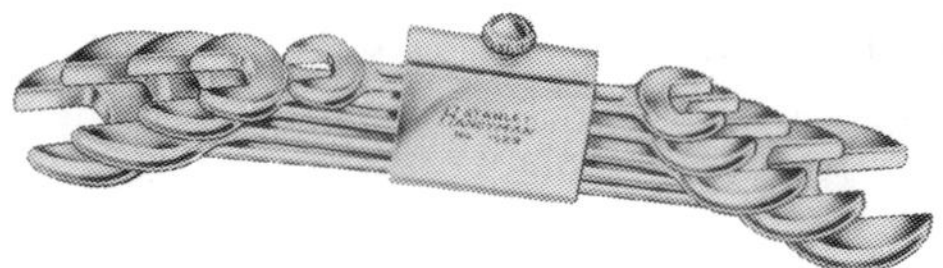

Fig. 8-54.

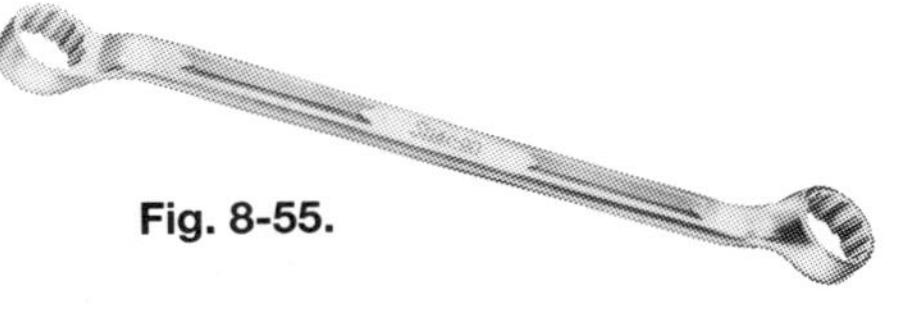

Fig. 8-55.

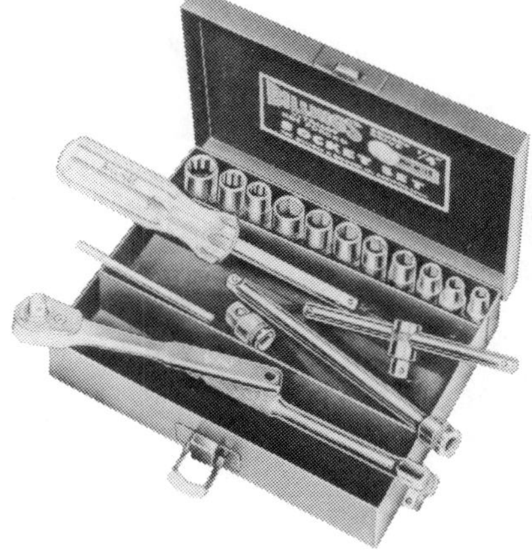

Fig. 8-56.

7. The tools here are numbered 8-51 through 8-56. Place these numbers in the blanks provided, matching the tools with their correct names, descriptions, and uses.

______ Used to assemble and disassemble machinery

______ Adjustable wrench

______ Uses replaceable metal-cutting blades

______ Cuts off a rivet or nail

______ Hacksaw

______ Makes adjustments on machines where there is plenty of clearance

______ A metal wrench with two enclosed ends

______ Socket wrench set

______ Used to install and replace knives and blades

______ Open-end wrench

______ Cuts all types of metal fasteners, hardware, and metal parts

______ Makes adjustments where there is limited space for movement

______ Fits many sizes of bolts and nuts

______ Cold chisel

______ Has U-shaped frame

______ Has a specially hardened and tempered cutting edge

______ A nonadjustable wrench

______ A series of sockets using a variety of handles

______ Box wrench

______ Will get a tight or rusted nut started

______ For variety of work, a complete set is needed

(Continued on next page)

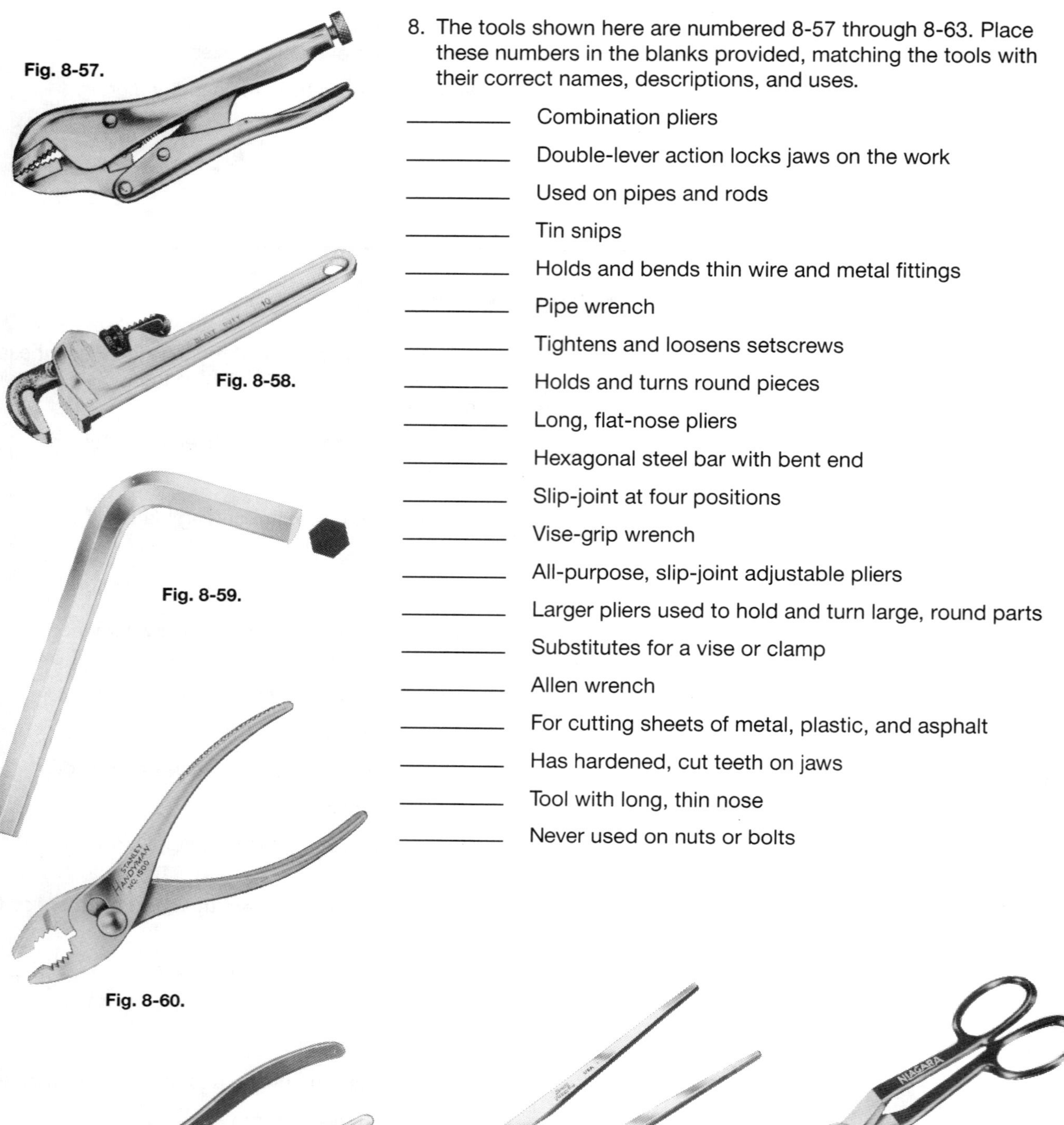

Fig. 8-57.

Fig. 8-58.

Fig. 8-59.

Fig. 8-60.

Fig. 8-61.

Fig. 8-62.

Fig. 8-63.

8. The tools shown here are numbered 8-57 through 8-63. Place these numbers in the blanks provided, matching the tools with their correct names, descriptions, and uses.

________ Combination pliers

________ Double-lever action locks jaws on the work

________ Used on pipes and rods

________ Tin snips

________ Holds and bends thin wire and metal fittings

________ Pipe wrench

________ Tightens and loosens setscrews

________ Holds and turns round pieces

________ Long, flat-nose pliers

________ Hexagonal steel bar with bent end

________ Slip-joint at four positions

________ Vise-grip wrench

________ All-purpose, slip-joint adjustable pliers

________ Larger pliers used to hold and turn large, round parts

________ Substitutes for a vise or clamp

________ Allen wrench

________ For cutting sheets of metal, plastic, and asphalt

________ Has hardened, cut teeth on jaws

________ Tool with long, thin nose

________ Never used on nuts or bolts

To be used with Unit 11 | Name ______________________________

Pages 122-127 | Score: (32 possible) ______

Study Unit 9

Portable Circular Saw

a. ____________________
b. ____________________
c. ____________________
d. ____________________
e. ____________________
f. ____________________
g. ____________________
h. ____________________

1. Name the parts of the portable circular saw shown in Fig. 9-1:

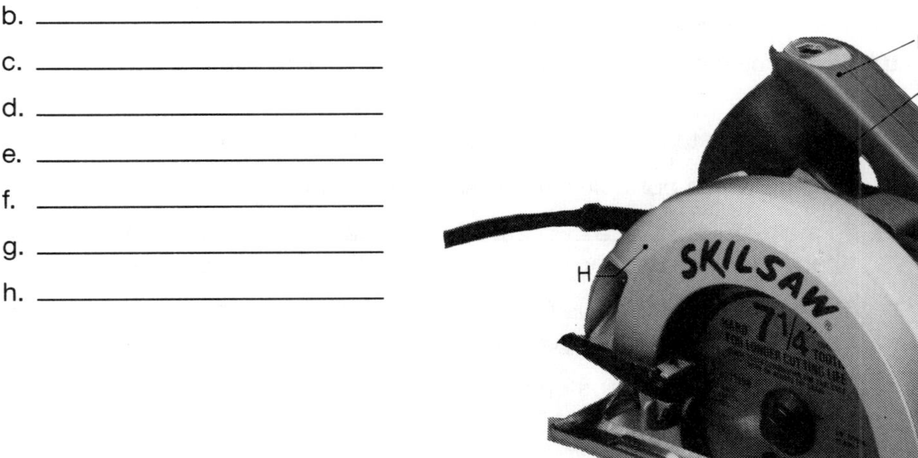

Fig. 9-1.

2. When using the portable circular saw:
   a. Make sure the blade ____________ are sharp and set correctly.
   b. Always keep the ____________ in place.
   c. Disconnect the ____________ source to change a blade.
   d. Allow the saw to reach ____________ speed before starting a cut.
   e. Keep your ____________ clear of the cutting line.
   f. Never stand in ____________ with the cut.
   g. Wait until the blade has ____________ before setting the saw down.

a. ____________________
b. ____________________
c. ____________________
d. ____________________
e. ____________________
f. ____________________
g. ____________________

____

3. In cutting plywood with a portable circular saw, the good side should be ____________.

____

4. Blades for a portable circular saw range in size from 6 to ____________ inches.

____

5. The size of a portable circular saw is determined primarily by the ____________ diameter, although the size of the motor is also important.

(Continued on next page)

_______________ 6. A protractor attachment for a portable circular saw is ideal for *(one wrong)*: a. angle cutting; b. miter cutting; c. compound angle cutting; d. circle cutting.

_______________ 7. The cutting action of the portable circular saw is similar to that of the circular saw. (T or F)

_______________ 8. The saw should be allowed to reach half speed before starting a cut. (T or F)

_______________ 9. The saw blade should clear the bottom of the work by about _______________ inch.

_______________ 10. An example of a pocket cut is the opening in a counter top for a sink. (T or F)

_______________ 11. It is possible to cut a miter freehand with the cutoff saw. (T or F)

_______________ 12. Ripping with a portable circular saw can be done freehand. (T or F)

_______________ 13. To make a _______________ cut with a portable circular saw, the shoe is tilted to the desired angle after loosening the wing nut or handle.

_______________ 14. The portable cutoff table for use with a portable circular saw can be used to cut flat miters but not edge miters. (T or F)

_______________ 15. _______________ miter stops are used as a guide when cutting flat miters on a cutoff table.

_______________ 16. _______________ miters are cut on the cutoff table by tilting the portable circular saw to the desired angle.

_______________ 17. A portable circular saw can be equipped with a dustbag. (T or F)

_______________ 18. Two types of circular saws are the _______________ saw and the worm-drive saw.

_______________ 19. A worm-drive saw has high blade speed and lower torque. (T or F)

To be used with Unit 12

Pages 128-140

Name ______________________

Score: (60 possible) ______

Study Unit 10
Table Saw

______ 1. Make all adjustments on the saw *(one right)*: a. when the power is off and the blade is motionless; b. while the power is on; c. during the cutting operation; d. just after the switch is turned off.

______ 2. Scraps should be removed from around the saw blade by *(one wrong)*: a. pushing the scraps away with a push stick; b. lifting the guard with one hand and removing the scraps with the other while the machine is running; c. removing the scraps after the machine has come to a dead stop; d. using a brush to remove the scraps.

______ 3. If the cut you are making does not permit the use of the regular guard, use the feather board or a ______ guard

______ 4. Always use a blade that is ______.

______ 5. Do not saw ______ material on the table saw.

______ 6. The size of the table saw is determined by *(one right)*: a. height of the saw; b. dimensions of the table; c. diameter of the recommended blade; d. size of the arbor.

______ 7. The table of a variety saw is permanently fastened in the horizontal position. (T or F)

______ 8. In selecting a saw blade, check the following *(one wrong)*: a. diameter of the blade; b. correct size of hole for arbor; c. kind of blade; d. brand name only.

9. Match the blades at the left with the descriptions on the right:

a. ______   a. ripsaw

b. ______   b. hollow ground

c. ______   c. plywood saw

d. ______   d. cutoff

e. ______   e. easy-cut

f. ______   f. combination saw

1. used primarily for trimming stock to length and squaring
2. has chisel-like teeth
3. used for a great variety of cutting
4. made of specially tempered steel
5. practically eliminates kickback, so considered to be the safest blade
6. used for fine cabinetwork

______ 10. A combination blade can be used for both ripping and crosscutting. (T or F)

______ 11. An accessory that will cut all widths of grooves is called a *(one right)*: a. dado head; b. molding head; c. cutter head; d. grooving head.

______ 12. The saw blade should protrude above the stock about *(one right)*: a. 1/4″; b. 1/8″; c. 1/2″; d. 1″.

(Continued on next page)

____

13. When removing a saw blade, remember the following about arbor threads *(one right)*: a. they may be either right-hand or left-hand threads, depending on the commercial make of the machine; b. they are left-hand threads, and you must turn the nut counterclockwise to loosen it; c. they are right-hand threads, and you must turn the nut clockwise to loosen it; d. they are double threads.

a. ______________________
b. ______________________
c. ______________________
d. ______________________
e. ______________________
f. ______________________
g. ______________________
h. ______________________
i. ______________________
j. ______________________
k. ______________________
l. ______________________
m. ______________________

14. Identify the parts of the saw shown in Fig. 10-1:

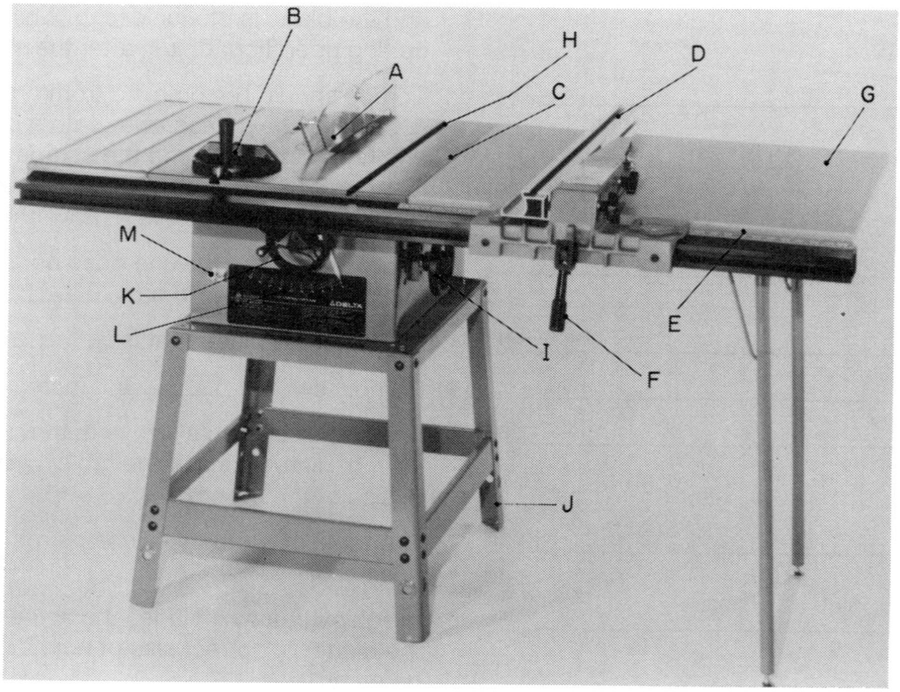

Fig. 10-1.

______________________ 15. For all ripping operations, use a ripping ______________ to guide the work.

______________________ 16. When ripping a wide board, apply pressure against the fence with the right hand and push the board forward with the left. (T or F)

______________________ 17. For ripping a board longer than 6′ to 8′, you should have a helper or a ______________ stand.

______________________ 18. When ripping narrower stock, the ______________ stick is used to take the place of your right hand.

______________________ 19. To rip extremely thick wood, cut partway through from opposite sides. (T or F)

______________________ 20. For crosscutting operations, the ______________ gauge is used.

______________________ 21. To prevent the stock from moving while the cut is made, use a ______________ rod.

______________________ 22. Plywood presents special cutting problems because of its construction. (T or F)

23. When cutting plywood, always place the stock on the table with the _____________ side up.

24. When cutting plywood, use one of the following techniques *(one wrong)*: a. reverse the miter gauge to start the cut; b. clamp a straightedge to the plywood; c. cut the plywood freehand; d. use the ripping fence as a guide.

25. When using a ripping fence to cut identical pieces to length, always use a _____________ block.

26. To cut a chamfer or bevel with the grain, the blade is tilted and the work held against the _____________.

27. Name the cuts and joint cuts shown in Fig. 10-2.

a. _____________
b. _____________
c. _____________
d. _____________
e. _____________

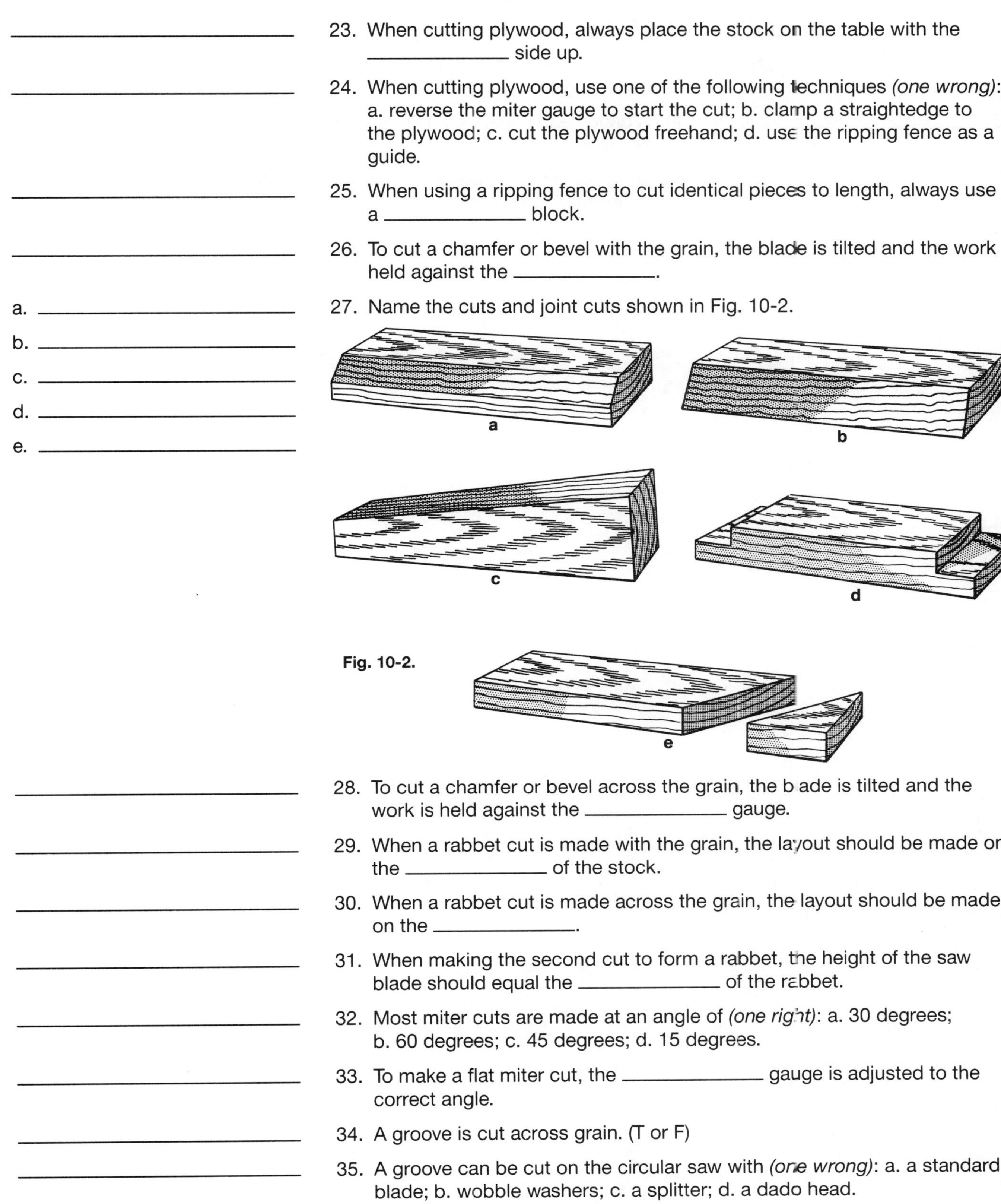

Fig. 10-2.

28. To cut a chamfer or bevel across the grain, the blade is tilted and the work is held against the _____________ gauge.

29. When a rabbet cut is made with the grain, the layout should be made on the _____________ of the stock.

30. When a rabbet cut is made across the grain, the layout should be made on the _____________.

31. When making the second cut to form a rabbet, the height of the saw blade should equal the _____________ of the rabbet.

32. Most miter cuts are made at an angle of *(one right)*: a. 30 degrees; b. 60 degrees; c. 45 degrees; d. 15 degrees.

33. To make a flat miter cut, the _____________ gauge is adjusted to the correct angle.

34. A groove is cut across grain. (T or F)

35. A groove can be cut on the circular saw with *(one wrong)*: a. a standard blade; b. wobble washers; c. a splitter; d. a dado head.

(Continued on next page)

_______________ 36. A dado is cut with the grain. (T or F)

_______________ 37. A dado cut only partway across the board is called a _______________ dado or gain.

_______________ 38. A dado head consists of two outside cutters with _______________ in between.

_______________ 39. Dado head cutters come in sizes which make it possible to cut a groove of any standard width. (T or F)

To be used with Unit 13

Pages 141-151

Name ____________________

Score: (41 possible) ______

Study Unit 11

Radial-Arm Saw

____________ 1. The radial-arm saw can be used for ripping, dadoing, grooving, and various combinations of these cuts. (T or F)

____________ 2. One advantage of the radial-arm saw is that a long board can be easily cut into shorter lengths by sliding the board across the table after cutting. (T or F)

____________ 3. When installing a saw blade, do the following *(one wrong)*: a. remove the wing nut to remove the guard; b. raise the blade so it will clear the table top; c. hold the blade with a block of wood; d. be sure the teeth at the bottom point away from you and toward the column.

____________ 4. The teeth of the saw blade should point in the direction of rotation. (T or F)

5. Name the parts of the radial-arm saw as shown in Fig. 11-1.

a. ____________
b. ____________
c. ____________
d. ____________
e. ____________
f. ____________
g. ____________
h. ____________
i. ____________
j. ____________
k. ____________
l. ____________
m. ____________
n. ____________
o. ____________
p. ____________
q. ____________
r. ____________

Fig. 11-1.

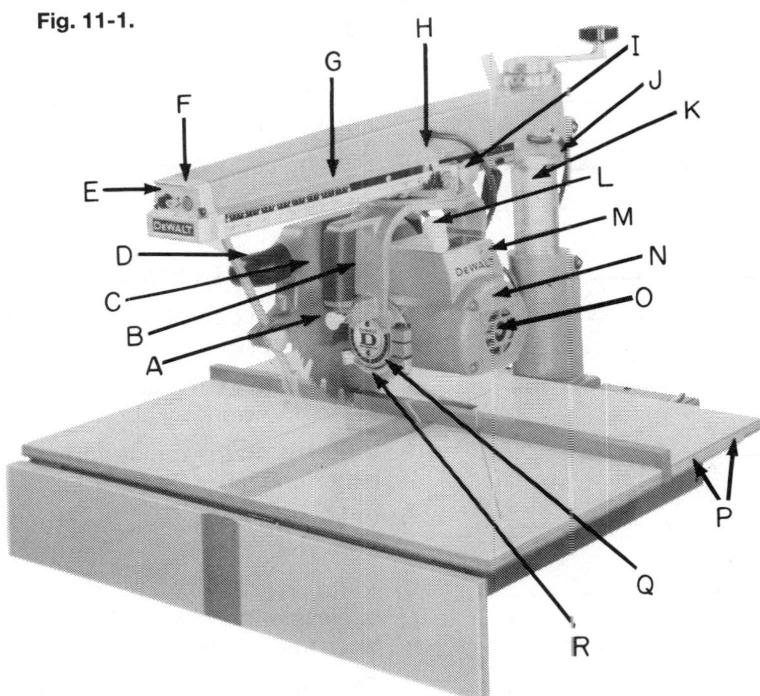

____________ 6. In crosscutting, the teeth of the saw should be about the following distance below the surface of the table *(one right)*: a. 1/32″; b. 1/16″; c. 1/8″; d. 1/4″.

(Continued on next page)

7. Crosscutting is done with the radial arm at ______________ angles to the guide fence.

8. Figure 11-2 shows the basic ways to ______________ on the radial-arm saw.

Fig. 11-2.

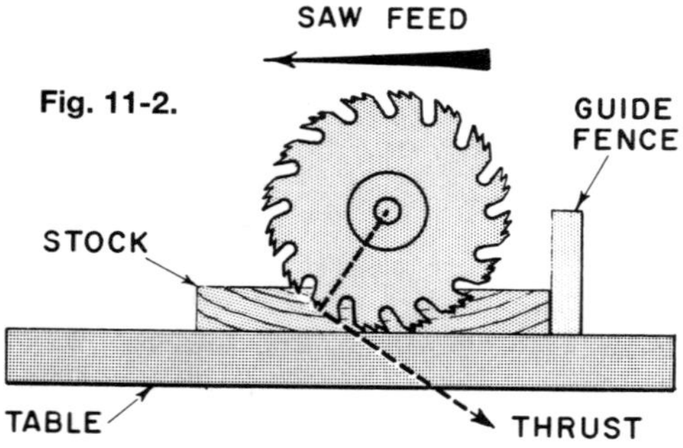

9. In crosscutting, the anti-kickback device should be about ______________ inch above the work surface.

10. For crosscutting, as well as mitering, dadoing, and similar operations, the work is held firmly to the table and against the ______________ fence.

11. Bevel cuts are made by tilting the ______________ to the right or left to the correct angle.

12. To make a compound miter cut, the following two must be adjusted *(one right)*: a. motor and yoke; b. motor and arm; c. arm and yoke; d. arm and table.

13. To cut a six-sided miter with the work tilted at 15 degrees, the blade must be set at ______________ degrees.

14. Ripping is done with the blade ______________ to the guide fence.

15. The motor is said to be ______________ when it is away from the column.

16. The motor is said to be ______________ when it is toward the column.

17. For ripping operations, the cutting head is stationary and the workpiece is moved. (T or F)

18. When ripping, the guard should be adjusted so that the infeed end clears the work by about ______________ inch.

19. When using the radial-arm saw for ripping, the stock can be fed from either end. (T or F)

20. When ripping, make sure that the blade is rotating downward and away from you. (T or F)

21. To rip an angle, you must *(one wrong)*: a. position the motor for ripping; b. elevate the saw; c. tilt the motor in the yoke; d. lower the saw until the teeth are 1/8″ below the table.

22. The dado heads used on a circular saw can also be used on the radial-arm saw. (T or F)

23. The saw teeth on a dado head should point the same way as with a saw blade. (T or F)

24. On most radial-arm saws *(one wrong)*: a. one revolution of the elevation crank lowers the blade 1/8″; b. three revolutions lower the blade 1/16″; c. two revolutions lower the blade 1/4″; d. four revolutions lower the blade 1/2″.

To be used with Unit 14

Pages 152-156

Name ______________________

Score: (29 possible) ______

Study Unit 12

Power Miter Saw

1. Follow these safety rules:
   a. Follow ____________ instructions.
   b. Do not disable the blade ____________.
   c. Make adjustments only after the ____________ has stopped moving.
   d. Before changing blades, ____________ the miter saw.
   e. Wear ____________ glasses.
   f. Wear ____________ protection as the saw has a high-pitched whine.
   g. Make sure the stock is ____________ supported by the saw table.
   h. The saw should be equipped with a blade-____________.

a. ______________________
b. ______________________
c. ______________________
d. ______________________
e. ______________________
f. ______________________
g. ______________________
h. ______________________

2. The power miter saw is sometimes called a power miter ____________.

____

3. Identify the parts of the power miter box shown in Fig. 12-1.

a. ______________________
b. ______________________
c. ______________________
d. ______________________
e. ______________________
f. ______________________

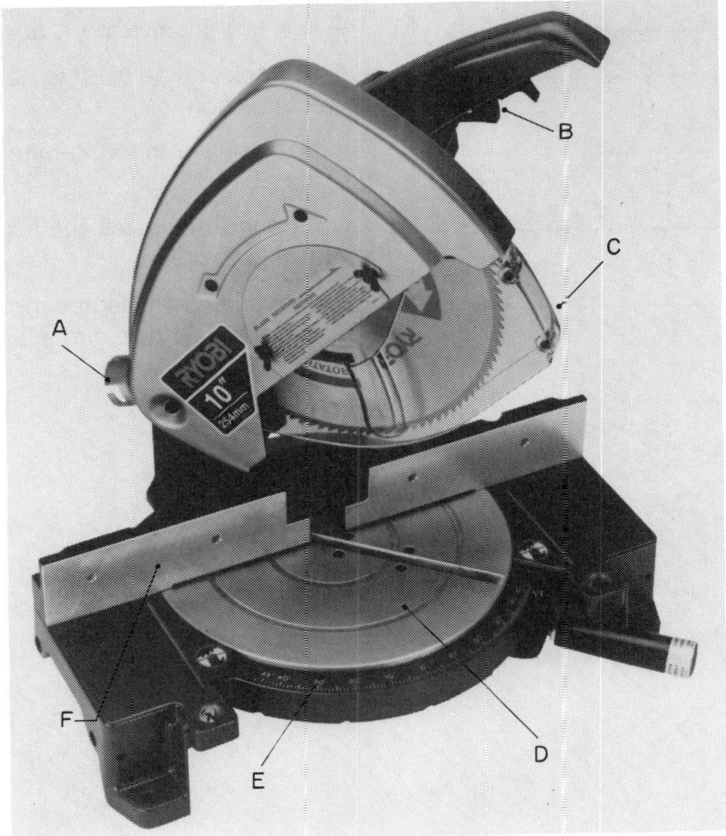

Fig. 12-1.

(Continued on next page)

4. The machine can cut miters either right or left. (T or F)

5. The power miter saw can be used for ripping. (T or F)

6. Saw blades range in size from 8″ to ______″.

7. The saw can be used to cut ______ extrusions.

8. A 2″ × 4″ can be cut at a 45-degree angle. (T or F)

9. When cutting flat pieces that are bowed, place the concave edge against the fence. (T or F)

10. With the correct kind of blade, the power miter saw can be used to cut the following *(one wrong)*: a. plastic pipe; b. aluminum extrusions; c. stainless steel; d. plywood.

11. When cutting aluminum, a ______ wax should be applied to the side of the blade.

12. The material positioned against the saw fence as in Fig. 12-2 is correct. (T or F)

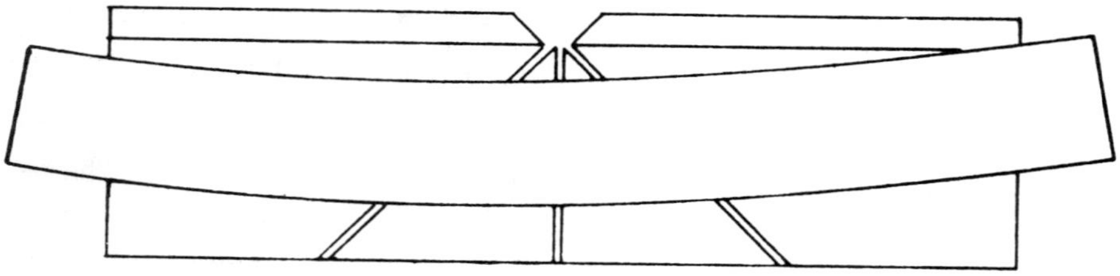

Fig. 12-2.

13. The saw has positive stops at 45 degrees and 90 degrees. (T or F)

14. A filler block can be used to hold the ______ molding in the correct position.

15. The blade of a compound-miter saw is mounted so that the blade can be ______.

16. A compound-miter saw can make a miter cut and a bevel cut at the same time. (T or F)

17. When mitering a 45-degree molding, the saw should be set at a 30-degree bevel angle and a ______-degree angle.

To be used with Unit 15

Pages 157-163

Name ______________________

Score: (35 possible) ______

Study Unit 13

Portable Electric Drill

______________ 1. Portable electric drills equipped with a variable speed motor can be used as a screwdriver by inserting a screwdriver bit. (T or F)

______________ 2. The most common sizes of electric hand drills are the 1/4″ and the ______________ inch.

______________ 3. When using a 1/4″, portable electric drill to hold a twist drill when drilling plastics, a high speed is used. (T or F)

______________ 4. To insert wood screws, the combination drill and countersink can be used. (T or F)

5. Name the parts of a 1/4″ portable drill as shown in Fig. 13-1.

a. ______________

b. ______________

c. ______________

d. ______________

e. ______________

f. ______________

g. ______________

h. ______________

i. ______________

j. ______________

k. ______________

l. ______________

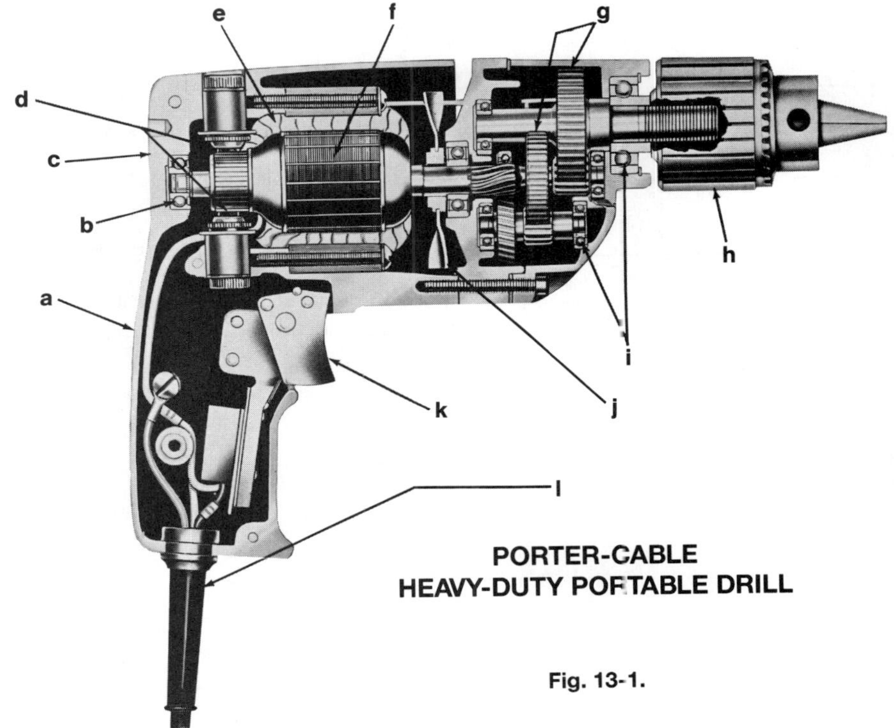

Fig. 13-1.

______________ 6. When driving screws with a portable electric drill, use a ______________ speed.

______________ 7. The combination drill will drill the pilot hole, shank hole, and ______________ in one operation.

______________ 8. The maximum head diameter of a number 7 screw is ______________ inches.

(Continued on next page)

__________ 9. The fractional-size drill that should be used for drilling the shank hole for a number-10 wood screw is __________ inch.

__________ 10. Plastic drill housings are called double-insulated housings. (T or F)

__________ 11. A cordless electric drill is powered by a __________ battery.

__________ 12. A portable tool used to drill holes in masonry materials is called a __________ drill.

__________ 13. Screw guns can be used to drill holes. (T or F)

__________ 14. Most twist drills are made of high-speed steel. (T or F)

__________ 15. A bit that has a large spur point and horizontal cutting surfaces is called a __________ bit.

__________ 16. Brad point bits have a center __________ that prevents the bit from "walking" as the hole is started.

__________ 17. Bits designed to cut deep holes quickly through wood are called __________ bits.

__________ 18. There are special bits for boring holes in glass, ceramic tile, and mirrors. (T or F)

__________ 19. Always __________ the drill or bit when changing cutting tools.

__________ 20. When drilling, too much pressure will make the bit dull. (T or F)

__________ 21. A Phillips screw bit is more difficult to use than a straight bit. (T or F)

__________ 22. A portable drill equipped with a variable speed control and reversing switch can be used as a powerful __________.

__________ 23. A number-24 wood screw has __________ threads per inch.

__________ 24. The nearest fractional equivalent of the root diameter of a number-12 wood screw is __________ inch.

To be used with Unit 16
Pages 164-173

Name ______________________

Score: (61 possible) ______

Study Unit 14

Pneumatic Nailers and Staplers

1. Complete the following safety precautions:
   a. Always wear __________ protection. Hearing protection is also __________.
   b. Never try to clear a __________ nailer while it is still connected to the air supply.
   c. Never fire a nailer unless the __________ is in contact with the workpiece.
   d. Never carry a nailer with your finger on the __________.
   e. Never use __________ gas to power the tool.
   f. Air-driven nails sometimes __________.
   g. Never operate a nailer at a __________ higher than it was designed to handle.
   h. If you are using a belt-driven compressor, make sure that the belts are protected by a __________.
   i. Before transporting a compressor, release the __________ in the tank.
   j. The __________ connected to a nailer should be in good condition.
   k. Make sure the nailer is pointing __________ when you connect a pressurized air hose to it.

a. __________
____
b. __________
c. __________
d. __________
e. __________
f. __________
g. __________
h. __________
i. __________
j. __________
k. __________

2. A pneumatic nailer or __________ is a tool that uses compressed air to drive fasteners into wood. __________
3. The two basic types of nailers are the __________ fed and the __________. __________ __________
4. Coil nailers are more common on job sites than strip nailers. __________
5. A strip nailer can be used longer without the need for reloading than a coil nailer. (T or F) __________
6. Strip nailers are easier to fit into tight places. (T or F) __________

(Continued on next page)

a. ______
b. ______
c. ______
d. ______
e. ______
f. ______

7. Name the parts of the nailer shown in Fig. 14-1.

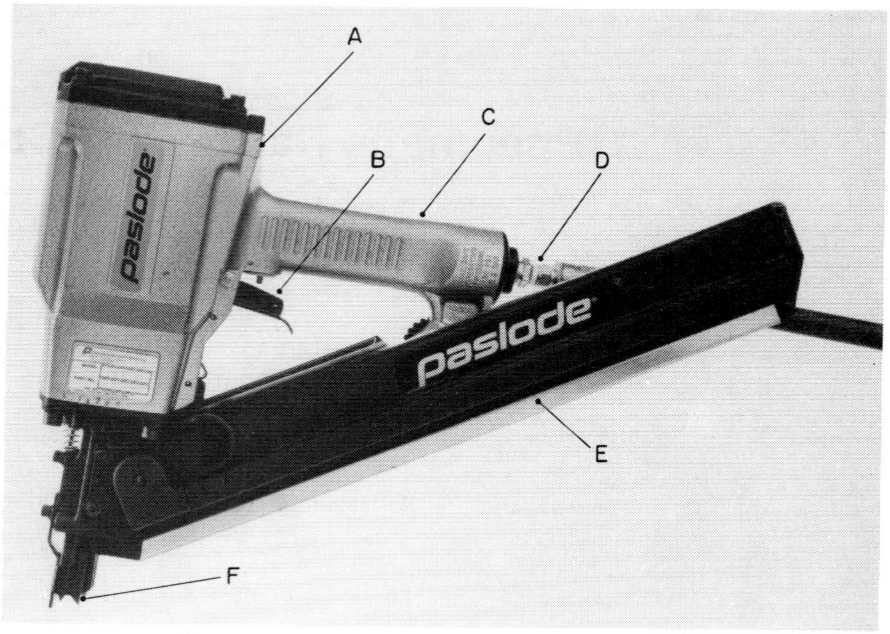

Fig. 14-1.

______ 8. Because of the narrow shape of the tool, coil nailers fit more easily into tight places. (T or F)

______ 9. All nailers and staplers employ a two-step ______ sequence.

______ 10. Nailers and staplers are attached to a hose that is linked to a ______.

11. The following two things must occur before a tool will fire:

a. ______ a. The ______ must be pulled.

b. ______ b. The ______ of the tool must be pressed against the workpiece.

______ 12. Most nailers operate on about ______ to ______ psi.

______

______ 13. A no-compressor nailer has a small internal-combustion engine located in the head of the tool. (T or F)

______ 14. Staplers are used primarily for installing *(one wrong)*: a. sheathing; b. sub-flooring; c. cabinets; d. roofing.

______ 15. Staples should be countersunk into the surface of the wood. (T or F)

______ 16. The crown of the staple should be kept perpendicular to the grain of the wood. (T or F)

______ 17. The crown width for staples used to install roofing must be at least ______ inch.

______ 18. Most building codes accept roofing staples on a one-for-one basis. (T or F)

______ 19. Staples used to install asphalt or fiberglass roofing must be ______.

_______________ 20. Staples for roofing should be placed _______________ inch above the shingle cutout.

_______________ 21. To maintain nailers and staplers in good condition, the tools must be properly _______________.

_______________ 22. Spray the _______________ of the tool with WD-40. Then wipe clean.

_______________ 23. Most compressors used on residential job sites are _______________ units.

_______________ 24. A valve that controls the amount of air pressure reaching a nailer is called a _______________.

_______________ 25. A line-pressure _______________ monitors the pressure in the hose leading to the air tool.

_______________ 26. If a worker can fire fasteners faster than a compressor can supply air, the compressor is _______________.

_______________ 27. An air hose should have a minimum inside diameter of _______________ inch.

_______________ 28. Long lengths of air hose can be a tripping hazard. (T or F)

_______________ 29. Air nailers are loaded with either strips or _______________ of nails.

_______________ 30. Three types of shanks on nails are *(one wrong)*: a. smooth; b. knurled; c. screw; d. ring.

_______________ 31. Nails range in length from _______________ inch brads to

_______________ _______________ inch spikes.

_______________ 32. Staples are available from _______________ inch long to _______________

_______________ inches long.

_______________ 33. Staples can be made from *(one wrong)*: a. steel; b. aluminum; c. copper; d. bronze.

_______________ 34. In framing, a 14-gauge 3-inch staple can be used instead of a _______________ nail.

_______________ 35. Staples used to install wall sheathing should penetrate the framing at least _______________ inch.

_______________ 36. Pneumatic tools can be used to install corrugated fasteners. (T or F)

_______________ 37. Staples used for framing must have a crown at least _______________ inch wide.

_______________ 38. Staples used for sheathing that is 1/2″ thick must be at least _______________ inches long.

_______________ 39. Corrugated fasteners are excellent for assembling picture frames. (T or F)

_______________ 40. The inside diameter for hoses 51 to 100 feet in length should be _______________ inch.

To be used with Unit 17

Pages 174-181

Name ______________________

Score: (51 possible) ______

Study Unit 15

Routers

1. When using the router:
   - a ______ a. Keep both ______ on the handles.
   - b. ______ b. Feed in the ______ direction.
   - c. ______ c. Always lay the router down with the cutter pointing ______ from you.
   - d. ______ d. Always hold onto the router when it is ______.
   - e. ______ e. Be certain the power switch is ______ before connecting the power plug.

______ 2. Routing is similar to ______ except that the motor and cutters are just opposite.

______ 3. The portable router consists of a motor with a ______ attached to the spindle.

______ 4. To do straight routing, a ______ is needed.

______ 5. A router can be used to cut a rabbet. (T or F)

______ 6. The shape of a decorative edge can be changed by moving the ______ up or down in the base.

7. Name the parts of the router as shown in Fig. 15-1.

a. ______
b. ______
c. ______
d. ______
e. ______
f. ______
g. ______
h. ______
i. ______
j. ______
k. ______
l. ______
m. ______
n. ______

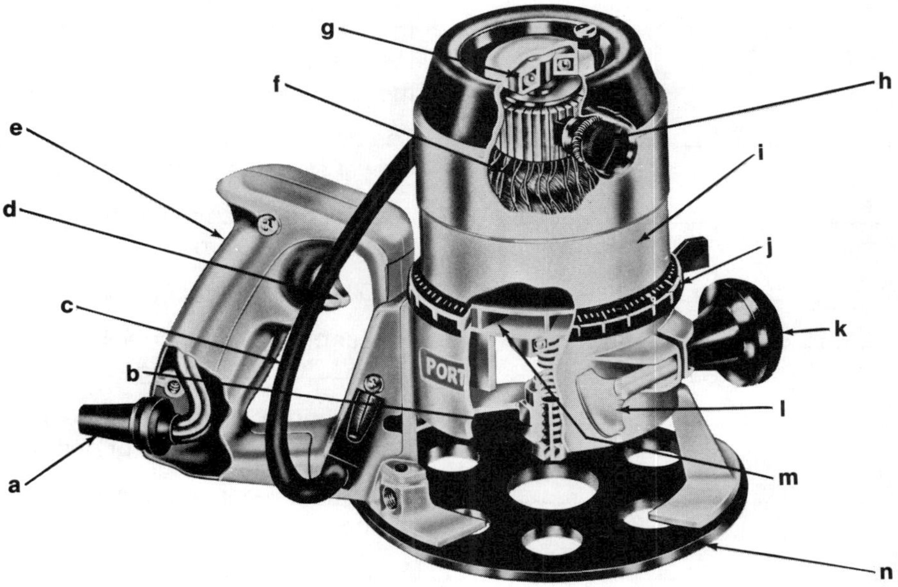

(Continued on next page)

a. ____________________
b. ____________________
c. ____________________
d. ____________________
e. ____________________
f. ____________________
g. ____________________
h. ____________________
i. ____________________
j. ____________________
k. ____________________
l. ____________________
m. ____________________

8. Name the router bits shown in Fig. 15-2.

Fig. 15-2.

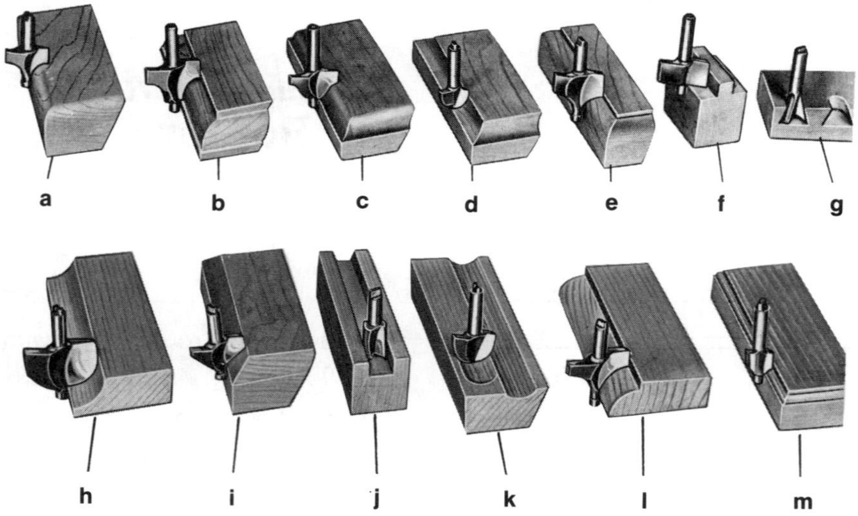

____________________ 9. Many kinds of decorative edges can be cut with a portable router. (T or F)

____________________ 10. When making a cut on a straight edge, feed from right to left. (T or F)

____________________ 11. When cutting on circular stock, feed in a ____________ direction.

____________________ 12. All router work is done freehand. (T or F)

____________________ 13. The joint used for good drawer construction is called the ____________ joint.

____________________ 14. The dovetail joint can be cut with a ____________ and dovetail attachment.

____________________ 15. In making a dovetail joint *(one wrong)*: a. clamp the dovetail attachment to a bench or table; b. clamp the project face side in against the front of the base; c. make a trial cut, being sure that the template guide follows the template; d. adjust for a shallower cut if the trail joint is too loose.

____________________ 16. The dovetail ____________ is a good joint to use for fastening a drawer side to a front when the front "lips over" the cabinet sides.

____________________ 17. A router on which the motor slides up and down on posts is called a ____________ router.

____________________ 18. A "D" handle router can be operated with one hand. (T or F)

____________________ 19. Router bits have either high-speed steel or ____________- ____________ cutting edges.

____________________ 20. Ball-bearing collars added to router bits can be used to change the ____________, width, or depth of the router bit.

____________________ 21. An adjustable circle-cutting jig works like a ____________ to guide the router.

____________________ 22. A router mounted in a router table operates much like a ____________.

To be used with Unit 18

Pages 182-185

Name ____________

Score: (28 possible) ______

Study Unit 16

Saber Saw

1. Follow these safety rules:
   a. Select the correct ________ for the work.
   b. Make sure the work is properly ________.
   c. Keep the cutting ________ constant.
   d. Hold the ________ securely on the work.

a. ________
b. ________
c. ________
d. ________

2. In making a heavy cut on a piece of 2″ × 4″ at a 45-degree angle, use a blade with ________ teeth per inch. ________

3. At least ________ teeth should be on the cutting surface at all times. ________

4. All saber saws are of the same design. (T or F) ________

5. Name the parts of the hand jigsaw shown in Fig. 16-1.

a. ________
b. ________
c. ________
d. ________
e. ________
f. ________
g. ________
h. ________
i. ________

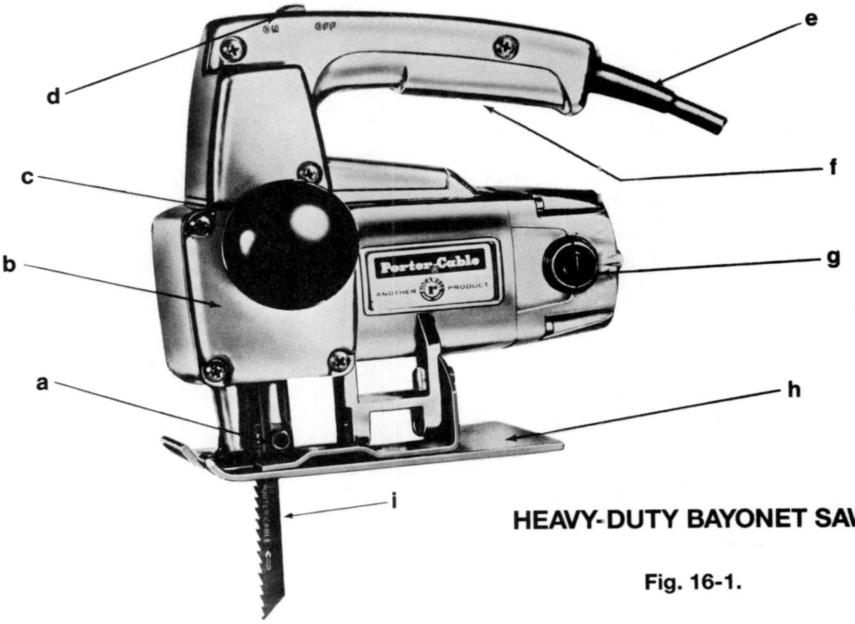

HEAVY-DUTY BAYONET SAW

Fig. 16-1.

6. The hand jigsaw is used primarily for straight and ________ cuts on the job. ________

7. Other names for the portable jigsaw are the ________ or bayonet saw. ________

8. Straight cutting can be done freehand or with the use of a ________ fence or guide. ________

9. Circles can be cut using the ripping ________ as a radius arm. ________

(Continued on next page)

_______________ 10. The correct blade to use for sawing cardboard is a _______________ blade.

_______________ 11. Making an inside cut without first drilling a hole is called _______________ cutting.

_______________ 12. For cutting plywood, a blade with the following number of teeth should be used *(one right)*: a. 6; b. 10; c. 12; d. 14.

_______________ 13. The hand jigsaw changes rotary action to up-and-down action. (T or F)

_______________ 14. A reciprocating saw operates with a back-and-forth movement. (T or F)

_______________ 15. A reciprocating saw is used mostly for _______________ or cabinetwork.

_______________ 16. When a reciprocating saw is used for remodeling, use a blade that will cut metal. (T or F)

_______________ 17. Blades for reciprocating saws are available that cut wood, plastic, metal, and _______________ rod.

To be used with Unit 19
Pages 186-189

Name ____________________

Score: (32 possible) ______

Study Unit 17

Portable Sander

1. Name the parts of the portable belt sander shown in Fig. 17-1.

a. ____________________
b. ____________________
c. ____________________
d. ____________________
e. ____________________
f. ____________________
g. ____________________
h. ____________________
i. ____________________
j. ____________________
k. ____________________
l. ____________________
m. ____________________
n. ____________________

Fig. 17-1.

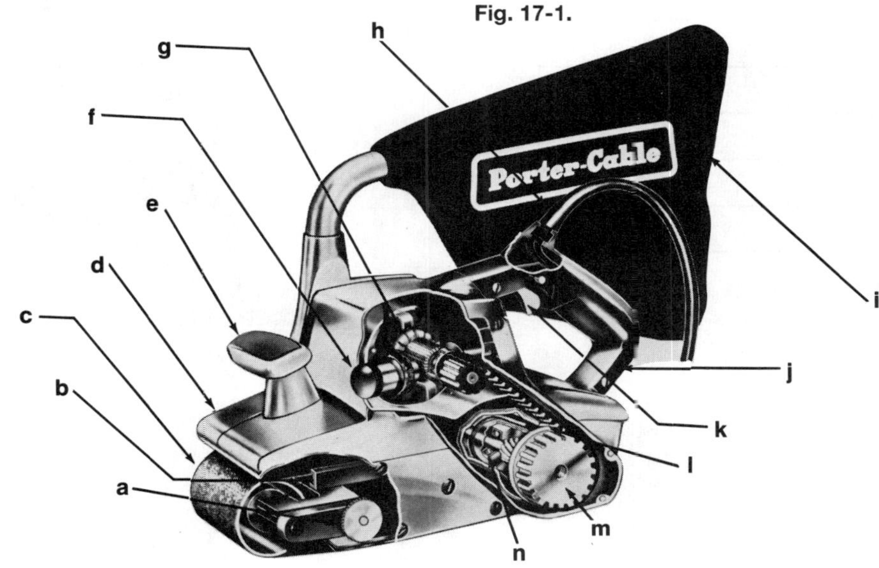

2. Identify the three principal methods of sanding with the finishing sander shown in Fig. 17-2.

a. ____________________
b. ____________________
c. ____________________

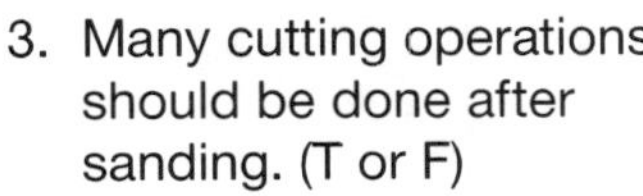

3. Many cutting operations should be done after sanding. (T or F) ____________________

4. To use a portable sander *(one wrong)*: a. put the cord over your right shoulder; b. hold the machine with your right hand; c. slowly lower the back of the sander onto the work; d. move the machine forward and back. ____________________

5. Finishing or pad sanders operate on one of three basic principles as shown in Fig. 17-2. Which action is least likely to leave cross-grain scratches: a, b, or c? ____________________

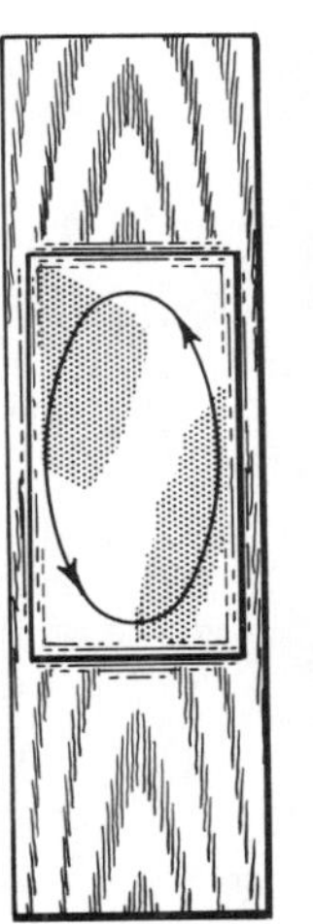

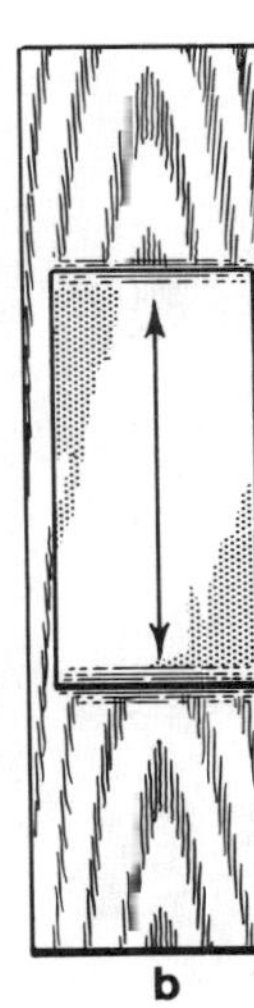

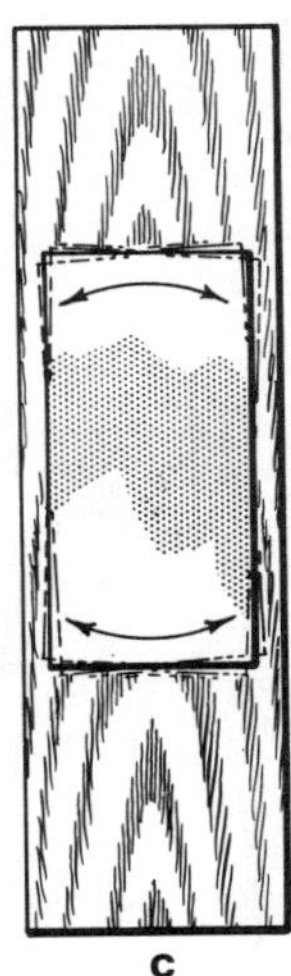

Fig. 17-2.

(Continued on next page)

______________________ 6. The finishing sander requires a great deal of hand pressure to do a good job. (T or F)

7. Name the parts of the finishing sander shown in Fig. 17-3.

a. ______________________

b. ______________________

c. ______________________

d. ______________________

e. ______________________

f. ______________________

g. ______________________

h. ______________________

i. ______________________

j. ______________________

k. ______________________

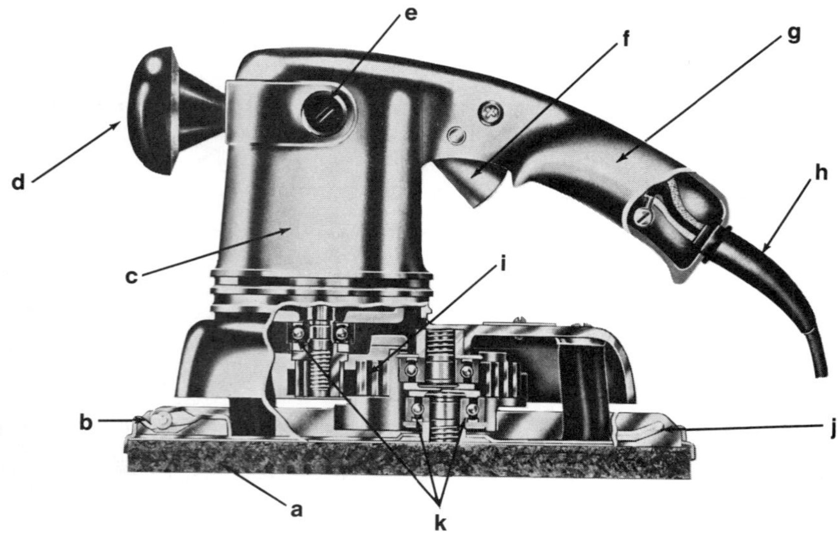

Fig. 17-3.

To be used with Unit 20 Name ______________________________

Pages 190-200 Score: (53 possible) ______

Study Unit 18

Jointers, Jobsite Planers, and Electric Planes

a. ______________________ 1. Name the jointer parts shown in Fig. 18-1.
b. ______________________
c. ______________________
d. ______________________
e. ______________________
f. ______________________
g. ______________________
h. ______________________
i. ______________________
j. ______________________
k. ______________________

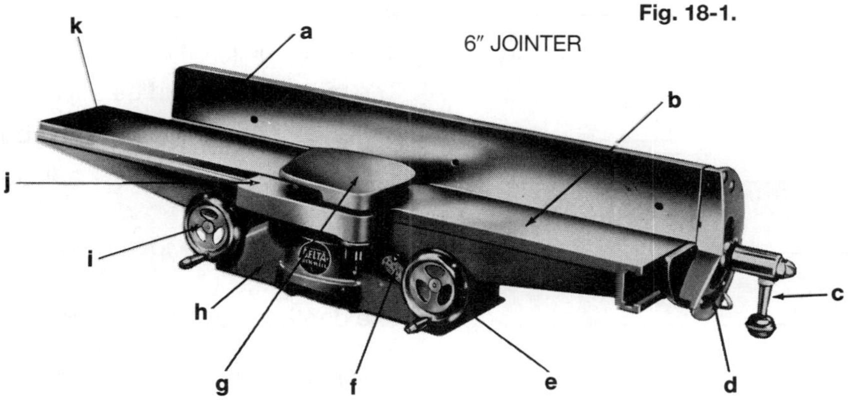

Fig. 18-1.

2. For the jointer to be safe to use, the following must be done:

a. ______________________ a. Keep the knives ____________.

b. ______________________ b. Be sure the fence is ____________.

c. ______________________ c. The guard must be in ____________ and operating easily.

d. ______________________ d. The machine should be at ____________ speed before it is used.

e. ______________________ e. Never stand ____________ the machine.

f. ______________________ f. The wood should be cut ____________ the grain.

g. ______________________ g. Keep the left hand back from the ____________ end of the board when feeding.

h. ______________________ h. Never apply pressure to the board with your hand directly over the ____________.

i. ______________________ i. Use good ____________ about when stock is too thin or short to joint safely.

______________________ 3. Operations that can be performed on the jointer include *(one wrong)*: a. planing a surface; b. cutting a rabbet; c. cutting a groove; d. cutting a bevel.

______________________ 4. The size of a jointer is indicated by the ____________ of the knives.

______________________ 5. A jointer should operate at a speed of about ____________ rpm.

______________________ 6. Another name for the outfeed table is ____________ table.

______________________ 7. Some jointers are made with one table that is adjustable and one table that is fixed. (T or F)

(Continued on next page)

8. Match the descriptions in the left column with the parts in the right column:

a. ______________________ | a. has three or four blades | 1. tables
b. ______________________ | b. covers the cutterhead | 2. cutterhead
c. ______________________ | c. infeed and outfeed | 3. fence
d. ______________________ | d. provides support for the work | 4. guard

9. The following adjustments must be made before work can begin:

a. ______________________ a. Align and adjust the ____________ table.

b. ______________________ b. Adjust the ____________ table.

c. ______________________ c. Adjust the position of the ____________.

______________________ 10. The rear table should be adjusted for each different type of cut. (T or F)

______________________ 11. If the rear table is too high, the board will be slightly tapered. (T or F)

______________________ 12. The outfeed table must be adjusted to change the depth of cut on a jointer. (T or F)

______________________ 13. It is usually a good idea to make a trial cut before beginning work. (T or F)

______________________ 14. The depth of cut depends on the following *(one wrong)*: a. amount of stock to be removed; b. whether the wood is soft or hard; c. whether the wood is open or closed grain; d. the smoothness of the surface desired.

______________________ 15. The position of the fence should always remain the same. (T or F)

______________________ 16. If a board has warp or wind, the cut must be made in a different way. (T or F)

______________________ 17. You always push the board across the jointer with the push ____________.

______________________ 18. The jointer can surface a board that is as wide as the knives. (T or F)

______________________ 19. When planing end grain, make a short cut at one end and then reverse the stock and feed from the opposite direction. (T or F)

______________________ 20. A guard can be used when planing end grain. (T or F)

______________________ 21. The jointer is a good machine to use when cutting a ____________ with the grain.

______________________ 22. When squaring up a board that is surfaced two sides (S2S), the edges are machined before the ends. (T or F)

______________________ 23. The width of the rabbet to be cut is controlled by the position of the *(one right)*: a. outfeed table; b. guard; c. fence; d. infeed table.

______________________ 24. The depth of the rabbet to be cut is controlled by the position of the *(one right)*: a. outfeed table; b. guard; c. fence; d. infeed table.

______________________ 25. The primary use of a jobsite planer is to surface the edges of boards. (T or F)

______________________ 26. The depth of cut of a jobsite planer should be set *(one right)*: a. while the machine is running; b. at any time; c. while the stock is being machined; d. when the machine is at a dead stop.

______________________ 27. The electric hand plane is rated by its horsepower size. (T or F)

______________________ 28. The depth of cut on an electric hand plane can be varied. (T or F)

______________________ 29. A power block plane can be held in one hand. (T or F)

______________________ 30. The electric hand plane can make bevel cuts. (T or F)

To be used with Unit 21 Name ______________________

Pages 201-204 Score: (43 possible) ______

Study Unit 19

Plate Joiners

1. Follow these general safety rules when using a plate joiner:
   a. Several practice cuts should be made on stock to ______________ the user with the machine. a. ______________
   b. A plate joiner ejects dust and ______________ at a high rate of speed. b. ______________
   c. Never cut a slot in a piece of wood that is ______________ held. c. ______________
   d. Wear ______________ protection because the plate joiner is noisy. d. ______________
   e. Failure to retract the blade fully may cause ______________. e. ______________
   f. Clamp the ______________ that is to be cut. f. ______________
   g. Chips and dust from the machine may cause a ______________ hazard. g. ______________
   h. Do not disable the ______________ points on the faceplate. h. ______________
   i. When changing the blades ______________ the power cord. i. ______________
   j. Make sure the blades are ______________. j. ______________
   k. Check the operation of the ______________ base. k. ______________
2. Another name for wood splines called plates is ______________. ______________
3. Biscuits strengthen the joint and help to ______________ the pieces accurately. ______________
4. The plate joiner cannot be used to install butt-joining custom wood flooring. (T or F) ______________
5. Name the parts of the plate joiner shown in Fig. 19-1.

a. ______________
b. ______________
c. ______________
d. ______________
e. ______________
f. ______________
g. ______________
h. ______________

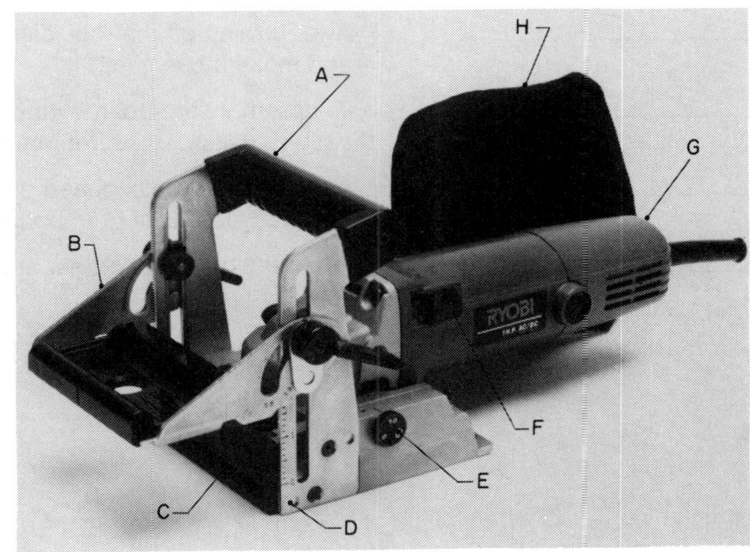

Fig. 19-1.

(Continued on next page)

______________________ 6. A plate joiner cuts ______________ shaped grooves in the edge of wood.

______________________ 7. The plates are die-cut from ______________ (wood) blanks.

______________________ 8. Plates or biscuits come in ______________ standard sizes.

______________________ 9. A number-20 plate or biscuit is about ______________ inch wide and ______________ inches long.

____

______________________ 10. Plates made of ______________ are used for joining synthetic counter-top materials.

______________________ 11. Wood plates absorb moisture from adhesives. (T or F)

12. Follow these steps in edge-joining two pieces of 1″ × 6″ board for shelves:

a. ______________________ a. Place the boards edge to ______________.

b. ______________________ b. Mark a line across the ______________ with a pencil.

c. ______________________ c. These marks should be 8 to ______________ inches on center.

d. ______________________ d. Adjust the joiner's depth of cut for the ______________ of the plate.

e. ______________________ e. Adjust the fence to ______________ the cut in the thickness of the board.

f. ______________________ f. Clamp ______________ board at a time to the workbench.

g. ______________________ g. Press the faceplate of the machine against the stock, using the centerline ______________ on the tool to align it with the layout marks.

h. ______________________ h. Turn the joiner on, and ______________ it into the board.

i. ______________________ i. After the cut is complete, pull the joiner ______________ from the stock.

j. ______________________ j. Line up the joiner with the next ______________ line. Continue to make cuts in this manner.

k. ______________________ k. After turning off the tool, clamp the ______________ piece in place and make those cuts.

l. ______________________ l. To assemble the boards, apply a light coat of glue to the inside of each ______________ on one board. Then insert the plates.

m. ______________________ m. Apply glue to the exposed portions of the plates and the ______________ of both boards.

n. ______________________ n. Press the boards together and ______________ them.

To be used with Unit 22 | Name ____________________

Pages 205-212 | Score: (29 possible) ______

Study Unit 20

Scaffolds and Ladders

a. ______
b. ______
c. ______
d. ______
e. ______
f. ______

1. Match the items at the left to the descriptions at the right:

a. bracket	1. constructed on the job to aid workers
b. manufactured scaffolding	2. support for scaffold parts
c. ladders	3. a pair of jacks that support scaffolding
d. wood scaffolding	4. can be set up or taken apart at building site
e. ladder jacks	5. usually made of wood or aluminum
f. trestle	6. can be attached to a ladder to hold a scaffold plank

______ 2. All scaffolding should be freestanding. (T or F)

______ 3. A ______ head nail is used when constructing wood scaffolding.

4. Scaffolding should have the following characteristics:

a. ______ a. Be plumb and ______.

b. ______ b. Be provided with proper guard ______.

c. ______ c. Have all ______ fastened securely.

d. ______ d. Should be climbed by means of ______ or fixed ladders.

e. ______ e. If freestanding, should be fastened by ______.

f. ______ f. Should rest firmly on the ______.

______ 5. The following are kinds of ladders *(one wrong)*: a. structural; b. straight; c. folding; d. extension.

6. Complete these statements on ladder safety:

a. ______ a. Never use ladders in a ______ position.

b. ______ b. Be sure that all ______ are tight.

c. ______ c. Check to see that ______ have a firm base.

d. ______ d. It is better to ______ the ladder than to lean far out to work.

e. ______ e. Remember that a metal ladder will conduct ______.

f. ______ f. Do not place a ladder in front of ______.

______ 7. When working on a roof, be sure the ladder extends above the roof edge at least ______ feet.

______ 8. An extension ladder should be leaned against a building at an angle of *(one right)*: a. 85°; b. 75°; c. 80°; d. 70°.

(Continued on next page)

_______________ 9. OSHA has developed regulations regarding safe construction techniques. (T or F)

_______________ 10. Improperly erected scaffolding is one of the leading causes of accidents in the construction industry. (T or F)

_______________ 11. When constructing wood scaffolding, use _______________ head nails.

_______________ 12. Horizontal wood pieces on which you stand are scaffolding _______________.

_______________ 13. Lumber plank is better than aluminum plank. (T or F)

_______________ 14. Laminated wood planks are made specifically for scaffolding. (T or F)

To be used with Unit 23

Pages 214-224

Name ____________________

Score: (25 possible) ______

Study Unit 21
Locating the House on the Building Site

______ 1. One way to locate a house on a lot is to work from a ______ point or line that can be identified.

______ 2. A street is a good reference point. (T or F)

______ 3. There are ______ ways to locate a structure on a lot. (T or F)

______ 4. Some reference points for laying out a site are *(one wrong)*: a. sidewalk; b. foundation of nearby building; c. a street light; d. a curbing.

______ 5. One instrument that can be used to locate a structure on a lot is the ______ level, also called the dumpy level.

______ 6. In addition to the dumpy level, the ______ level can be used to locate a structure on a lot.

______ 7. A reference point is not needed when using one of the levels in questions 5 and 6. (T or F)

Fig. 21-1.

______ 8. The instrument shown in Fig. 21-1 is an ______ level.

______ 9. The instrument shown in Fig. 21-2 is a ______ level.

______ 10. The instrument used for measuring horizontal angles only is the ______ level.

Fig. 21-2.

______ 11. By careful use of the level, one can see the difference in ______ between two points.

______ 12. To lay out an irregularly shaped structure on a lot, more points are needed than for a rectangular shape. (T or F)

______ 13. A reference point which refers to the level of the ground where it will touch the foundation of the completed building is called the ______ line.

(Continued on next page)

______________ 14. When setting up a level, the point over which the level is directly centered is called the ______________ mark.

______________ 15. Once a level has been set up, it can be easily moved to adjust the position of the tripod. (T or F)

______________ 16. After the corners of the house have been located on the lot, the outline of the house is marked with *(one right)*: a. stakes; b. a chalk line; c. metal posts; d. batter boards.

______________ 17. The ground level of a lot after the lot has been graded is called the ______________ grade.

______________ 18. Foundation walls should extend above this ground level *(one wrong)*: a. for better appearance; b. to protect the wood finish of the house; c. to protect the house frame; d. to keep the house above the grass line.

______________ 19. Test borings of subsoil should be taken to determine *(one wrong)*: a. if oil is present; b. the height of the water table; c. if there is a rock ledge; d. the kind of soil.

______________ 20. Footings and basement should be excavated at the same time. (T or F)

______________ 21. Basement excavations should always have vertical walls. (T or F)

______________ 22. The outside of the foundation wall is the line for the basement excavation. (T or F)

______________ 23. The telescope of a ______________ is fixed in position.

______________ 24. A single crew member can use a ______________ level.

______________ 25. One type of transit, called an electronic transit, reads ______________ electronically.

To be used with Unit 24 Name ______________________

Pages 225-235 Score: (33 possible) ______

Study Unit 22

Concrete and Footings

______ 1. Concrete contains *(one wrong)*: a. gravel; b. water; c. a chemical hardening agent; d. sand.

______ 2. When water is added to concrete ingredients, a reaction called ______ takes place.

______ 3. This reaction causes concrete to ______.

______ 4. Water used for making concrete should be clean enough to drink. (T or F)

______ 5. Clay mixed with gravel helps to strengthen the concrete. (T or F)

______ 6. The proportions for mixing concrete are relatively unimportant. (T or F)

7. A 1:2:3 batch of concrete has the following proportions:

a. ______ a. One part ______.

b. ______ b. Two parts ______.

c. ______ c. Three parts ______.

______ 8. Whenever possible, pour concrete continuously. (T or F)

______ 9. The area where concrete is being poured should be kept level. (T or F)

______ 10. Spade or ______ the concrete to remove air pockets.

______ 11. Concrete should be allowed to dry rapidly. (T or F)

______ 12. The faster concrete dries, the stronger it is. (T or F)

______ 13. Concrete sets at a rate directly related to temperature. (T or F)

______ 14. Footings are the ______ of a structure.

______ 15. Footings are a standard size regardless of climate or soil condition. (T or F)

______ 16. The depth of footings varies with the climate. (T or F)

______ 17. The best footings for homes are made of concrete. (T or F)

______ 18. Footings should be at least *(one right)*: a. 6″ thick; b. 7″ thick; c. 5″ thick; d. 8″ thick.

19. Two rules for footings are:

a. ______ a. Footing ______ should equal wall thickness.

b. ______ b. Footings should project out from sidewalls ______ the thickness of the walls.

______ 20. Footings should go below the frost line. (T or F)

______ 21. Soil condition has little to do with footing size. (T or F)

(Continued on next page)

_______________ 22. Footings are required for *(one wrong)*: a. chimneys; b. fireplaces; c. furnaces; d. sidewalks.

_______________ 23. Piers, posts, and columns are usually made in two sizes: 24″ × 24″ × 12″ and _______________.

_______________ 24. When a lot is steep, the footings are poured in the shape of _______________.

_______________ 25. To help keep moisture out of a home, foundation or footing _______________ are installed.

26. Match the admixtures on the left with the descriptions on the right.

a. _______________
b. _______________
c. _______________
d. _______________
e. _______________

a. air-entraining
b. retarding
c. accelerating
d. water-reducing
e. superplasticizing

1. concrete sets up slowly
2. can do one of two things
3. makes the concrete stronger
4. introduces bubbles in the concrete
5. concrete can be put in service quickly

To be used with Unit 25 | Name ______________________

Pages 236-243 | Score: (33 possible) ______

Study Unit 23

Poured Concrete Foundation Walls

______ 1. In addition to soil conditions, building ______ must also be considered when planning construction of foundation walls.

______ 2. Foundation walls bear the weight of an entire house. (T or F)

______ 3. Basement wall height should be at least *(one right)*: a. 7′; b. 6 1/2′; c. 8′; d. 8 1/2′.

______ 4. For a poured concrete wall, each side of the wall must have a wood form. (T or F)

______ 5. This wood form is called ______.

______ 6. Forms can be used only once. (T or F)

______ 7. If the forms are straight and plumb, bracing is unnecessary. (T or F)

______ 8. During good weather, poured foundation walls dry in not less than *(one right)*: a. 24 hours; b. three days; c. a week; d. two days.

______ 9. To keep poured concrete walls from rain seepage, they should be ______.

______ 10. Concrete walls are protected from rain seepage by an application of hot tar or ______.

______ 11. A house with a basement is more expensive than one built with a crawl space. (T or F)

______ 12. In a house with a crawl space, the floor beams are supported by ______ made of concrete or concrete block.

______ 13. These should extend above the ground line at least *(one right)*: a. 10″; b. 12″; c. 14″; d. 16″.

______ 14. The size of exterior wall piers depends on the weight of the house. (T or F)

______ 15. The brick should be completely sealed with mortar. (T or F)

______ 16. The device which gives strength and stability to the sill plate is the ______.

______ 17. The device named in question 16 must be imbedded deeper in concrete block than in poured concrete walls. (T or F)

______ 18. Anchor bolts should be placed ______ feet apart.

______ 19. Poured concrete walls must have reinforcements over doors and windows called ______.

______ 20. Wood-frame walls covered with a masonry veneer such as brick must be tied to the veneer with metal ______.

______ 21. Sheathing should be applied between brick and wood-frame walls. (T or F)

(Continued on next page)

a. ______________________

b. ______________________

c. ______________________

d. ______________________

e. ______________________

f. ______________________

22. Name the construction parts shown in Fig. 23-1.

Fig. 23-1.

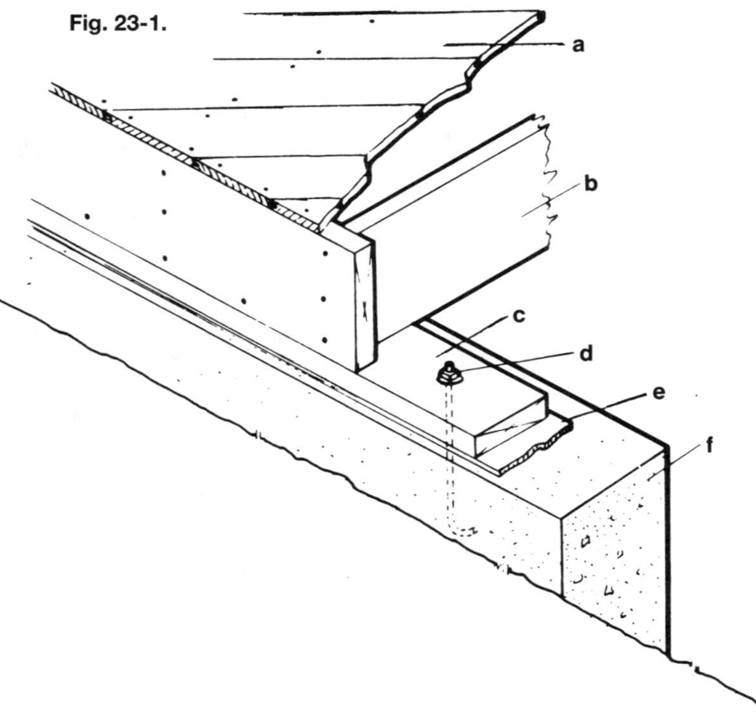

______________________ 23. Ventilation is needed when a wood beam is laid on a concrete wall to prevent *(one right)*: a. warp; b. wind; c. rust; d. decay.

______________________ 24. Termite damage occurs uniformly throughout the United States. (T or F)

______________________ 25. One of the best ways to control termites is to treat the ______________.

______________________ 26. Using ______________ to control termites is one of the most common and effective methods.

______________________ 27. The amount of concrete required for the 12″ thick and 8′ high foundation walls in Fig. 23-2 would be about *(one right)*: a. 39 cu. yd.; b. 26 cu. yd.; c. 30 cu. yd.; d. 45 cu. yd.

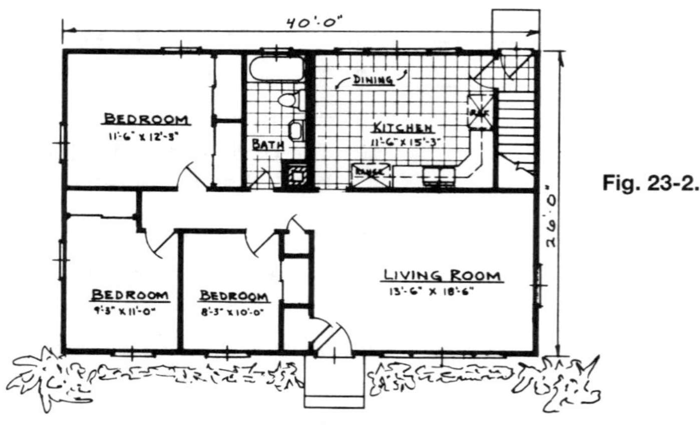

Fig. 23-2.

______________________ 28. It takes an average of ______________ hours to place 1 cubic yard of concrete in the forms for a wall.

To be used with Unit 26 | Name ____________________

Pages 244-262 | Score: (42 possible) ______

Study Unit 24

Concrete Block Foundation Walls

______ 1. Modular concrete blocks are usually 7 5/8 inches high and ______ inches long.

______ 2. Concrete block comes in one width or thickness. (T or F)

______ 3. A series of building materials installed one layer over the other is called a ______.

______ 4. The height of a basement wall made of concrete blocks is about ______ as measured from the joists to the basement floor.

______ 5. The number of courses customarily laid to build a concrete block wall for a basement is ______.

______ 6. Laying concrete block requires the use of forms. (T or F)

______ 7. The mortar joints between concrete blocks should measure in width about *(one right)*: a. 1/4″; b. 3/8″; c. 1/3″; d. 1/2″.

______ 8. A concrete block wall may be reinforced by a column of blocks called a ______.

______ 9. In the stack bond pattern of laying cement block, the center of one block is laid over the joint in the two below. (T or F)

______ 10. Reinforcement is required for the stack bond pattern. (T or F)

______ 11. Freezing weather has the following effects on mortar *(one wrong)*: a. faster setting; b. low adhesion; c. less strength; d. failure in joints.

______ 12. Waterproofing with a coat of ______ after applying a coating of cement-mortar will normally assure a dry basement.

______ 13. Extra waterproofing can be provided by applying roofing ______ or a similar material over the first waterproof coating and covering this with hot tar or asphalt if desired.

14. Mortar strength depends upon:

a. ______ a. The rate at which concrete block absorbs ______ from the mortar.

b. ______ b. The plasticity or ______ of the mortar.

c. ______ c. How much ______ the mortar retains.

d. ______ d. The quality of ______ displayed by the mason.

______ 15. When air temperature is 80°F. or higher, mortar will remain usable for *(one right)*: a. 1 hour; b. 3 hours; c. 2 1/2 hours; d. 3 1/2 hours.

______ 16. Concrete block should be watered down before it is used. (T or F)

______ 17. Several of the first blocks should be positioned before mortar is applied. (T or F)

(Continued on next page)

_______________ 18. The _______________ block should be the first one laid.

_______________ 19. Proper alignment of the first course of blocks is less important than for succeeding courses. (T or F)

_______________ 20. A larger mortar-bedding area is provided by laying the blocks with the smaller holes down. (T or F)

_______________ 21. To make sure the blocks are laid in an even line, check each block with a _______________ or straightedge.

_______________ 22. Use a _______________ pole or a course pole to make sure the top of each block is 8″ above the previous one.

_______________ 23. The final gap in the wall is filled with a _______________ block.

_______________ 24. Pressing your thumb into mortar leaves an imprint if the mortar is "thumbprint hard." (T or F)

_______________ 25. A process of pressing mortar tightly into the joints on both sides is called _______________.

_______________ 26. Tooling should be done while the mortar is still very soft. (T or F)

_______________ 27. A control joint is built into a concrete block wall for the following reasons *(one wrong)*: a. to control cracking due to stress; b. to permit slight movement in the wall; c. to give the wall longer life; d. to make the wall waterproof.

_______________ 28. A control joint *(one wrong)*: a. should be thicker than the mortar joints; b. is built vertically into the wall; c. should be plumb; d. should be caulked if it shows.

_______________ 29. To prevent mortar from smearing on concrete block, remove the mortar spatters after they have dried. (T or F)

_______________ 30. It is not necessary to anchor intersecting concrete block walls. (T or F)

_______________ 31. The tops of windows and doors require lintels. (T or F)

_______________ 32. Fiberglass-reinforced mortar should be mixed for 10 to 15 minutes. (T or F)

_______________ 33. When applying fiberglass-reinforced mortar, a sheet metal device called a _______________ guide is used when the application is interrupted for more than an hour.

_______________ 34. The wall should be wetted down prior to fiberglass mortar mix application. (T or F)

_______________ 35. The finished grade should be sloped away from the foundation walls for good surface drainage. (T or F)

_______________ 36. Backfill material should be expected to settle. (T or F)

_______________ 37. Less block is needed per square foot of wall area with fiberglass-reinforced mortar than is needed for conventional mortared block walls. (T or F)

_______________ 38. Split-face blocks have one rough face that looks something like _______________.

_______________ 39. There are several kinds of specialty blocks for specific purposes. (T or F)

To be used with Unit 27

Pages 263-274

Name ____________________

Score: (43 possible) ______

Study Unit 25

Slab and Flatwork

____________ 1. Concrete flatwork is usually ____________ inches or less in thickness.

____________ 2. Flatwork construction is the best foundation choice for a house without a basement. (T or F)

____________ 3. In a combined slab and foundation the footing should extend below the grade line *(one right)*: a. 6″; b. 8″; c. 10″; d. 12″.

____________ 4. The slab requires a vapor barrier. (T or F)

____________ 5. In cold climates, foundations must extend below the ____________ line.

____________ 6. Good vapor barriers include *(one wrong)*: a. heavy plastic film; b. asphalt-laminated sheet; c. gravel; d. heavy roofing.

7. Concrete floor slabs require certain construction details:

a. ____________ a. Vapor barrier laid under the ____________.

b. ____________ b. Finish floor level above the natural ____________ level.

c. ____________ c. Removal of ____________.

d. ____________ d. After laying of water and sewer lines, a covering of ____________ or crushed rock.

e. ____________ e. A good rigid insulation installed around the outside of the ____________.

f. ____________ f. A smooth, troweled ____________.

____________ 8. No reinforcement is required for concrete floor slabs. (T or F)

____________ 9. A floor slab will not settle uniformly unless the earth below the slab has been ____________.

____________ 10. Before tamping it down, the subgrade should *(one wrong)*: a. be thoroughly dampened; b. have all organic material removed; c. be rough graded; d. be leveled off.

____________ 11. After the subgrade has been prepared, a layer of ____________ is compacted over it.

____________ 12. Fill may consist of *(one wrong)*: a. crushed stone; b. coarse slab; c. wood chips; d. gravel.

____________ 13. The granular layer is then covered with a stiff coat of ____________.

____________ 14. After applying a dampproofing membrane, the concrete floor slab is poured. (T or F)

____________ 15. Compacting the cement for the floor slab can be done by *(one wrong)*: a. vibrating; b. spading; c. tamping; d. raking.

(Continued on next page)

______ 16. Concrete should be allowed to cure *(one wrong)*: a. by natural drying in the air; b. covered with burlap; c. by sprinkling with water to keep the covering wet; d. covered with waterproof paper.

______ 17. Linoleum and asphalt cannot be installed over concrete unless the concrete surface has been smoothed with a steel ______.

______ 18. One disadvantage to using concrete as flooring is that it can't be colored. (T or F)

19. Match the items at the left with the descriptions at the right:

a. ______ b. ______ c. ______

a. porches and basement floors	1. surface should be coarse
b. driveways	2. surface should have smooth-troweled finish
c. sidewalks	3. surface should be non-slippery

______ 20. The process of removing excess concrete is called ______.

______ 21. Excess concrete is removed in order to make the surface ______.

______ 22. Concrete is sometimes edged to prevent ______ or damage.

______ 23. Random cracks that develop in concrete can be controlled by cutting contraction ______ in the surface.

______ 24. The tool shown in Fig. 25-1 is a wood ______.

Fig. 25-1.

______ 25. For a smoother finish, screeding is followed by ______.

______ 26. If done too soon, floating raises fines and ______ to the surface of concrete.

______ 27. The tool shown in Fig. 25-2 is a metal ______.

______ 28. The tool in Fig. 25-2 makes the concrete surface *(one wrong)*: a. smooth; b. rippled; c. free of marks; d. even.

Fig. 25-2.

______ 29. The cement finisher can work on large areas without leaving dents by using ______.

______ 30. A 12 percent grade in a driveway means that it rises 12 feet in ______ feet.

______ 31. Two basic types of driveway are the slab and the ______ types.

______ 32. Concrete for a driveway should be about ______ inches thick.

______ 33. Expansion joints in a driveway should be located *(one wrong)*: a. where driveway meets a sidewalk; b. along a curb; c. about 40′ apart; d. about 20′ apart.

______ 34. Sidewalks that slope should have a grade not greater than ______ percent.

______ 35. The basement floor should be equipped with at least one ______.

______ 36. A basement floor requires no screeding. (T or F)

To be used with Unit 28

Pages 276-284

Name ______________________________

Score: (28 possible) ______

Study Unit 26

Framing Methods

______ 1. Wood-frame construction is found in most homes on this continent. (T or F)

2. A wood-frame house has the following characteristics:

a. ______ a. Exterior of stucco, brick, wood siding, or wood ______.

b. ______ b. Costs ______ than other types.

c. ______ c. Insulating qualities are ______ than other types.

d. ______ d. Almost any ______ style can be produced.

______ 3. A disadvantage of a wood-frame house is its lack of durability. (T or F)

______ 4. Platform and balloon are two kinds of ______ framing.

______ 5. The construction method most commonly found in one-story houses is platform-frame construction. (T or F)

______ 6. A house frame with lumber and covered with brick or stucco is considered a wood-frame house. (T or F)

______ 7. In platform-frame construction *(one wrong)*: a. the building process is easier than with balloon frame; b. each floor is constructed separately; c. wall framing can be assembled on the floor and tilted into place; d. the cost is higher than balloon framing.

______ 8. In balloon framing, studs and first-floor ______ rest on the foundation.

______ 9. Balloon framing is less affected by expansion and contraction than platform framing. (T or F)

______ 10. Post-and-beam construction is not one of the conventional framing methods. (T or F)

______ 11. Post-and-beam construction *(one wrong)*: a. has interior roof planks exposed; b. has smaller framing members; c. lends itself to striking architectural effects; d. requires less labor.

______ 12. The framing pieces for metal framing are made in a factory. (T or F)

______ 13. Paul Revere's house is older than the "House of the Seven Gables." (T or F)

______ 14. Metal framing has the following advantages *(one wrong)*: a. will not rot or decay; b. does not warp or twist; c. will not expand with temperature; d. will not shrink or swell with changes in temperature.

______ 15. Metal framing can be produced in lengths not available with wood framing. (T or F)

______ 16. The top and bottom plates of metal framing are called ______.

(Continued on next page)

_______________ 17. The screws used to assemble metal framing have Phillips heads. (T or F)

_______________ 18. Metal framing for exterior walls is common in home construction. (T or F)

_______________ 19. Windows for metal framing are prefabricated and painted. (T or F)

_______________ 20. Wiring for metal framing passes through insulated _______________.

_______________ 21. Panels of _______________ are sometimes used to form the shell of a house.

_______________ 22. The panels consist of thick rigid insulation that is _______________ inch thick sandwiched between sheets of plywood or OSB board.

_______________ 23. A timber frame is a structural skeleton of posts and _______________ connected with wooden joinery.

_______________ 24. A timber frame is a freestanding structural system. (T or F)

_______________ 25. Interlocking wood joinery for timber framing is secured by wooden _______________.

To be used with Unit 29
Pages 285-315

Name ______________________

Score: (39 possible) ______

Study Unit 27

Floor Framing

1. Match the floor framing parts at the left with the descriptions at the right:

a. ______
b. ______
c. ______
d. ______
e. ______

a. joists
b. posts
c. sills
d. girders
e. subflooring

1. wood laid over floor joists
2. pieces of wood or steel which support the girders
3. large beams which support floor joists
4. rest on top of girders
5. anchored to the foundation wall for fastening and supporting the joists

2. Name the parts shown in Fig. 27-1.

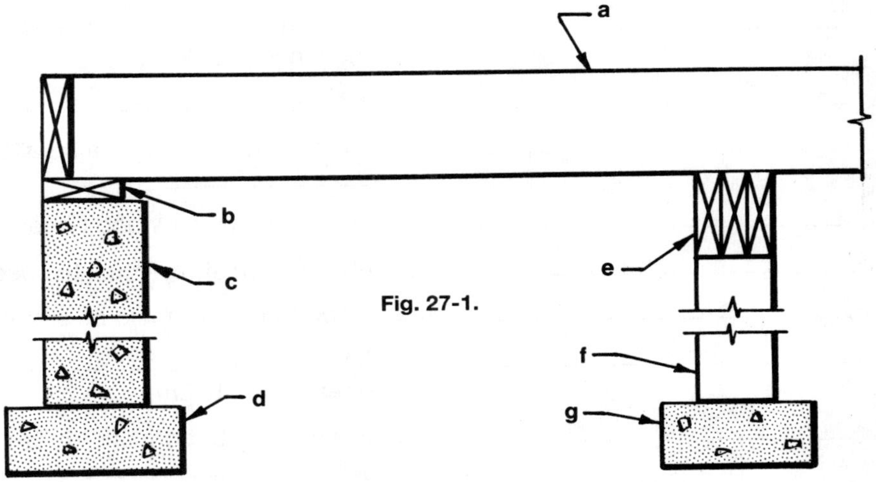

a. ______
b. ______
c. ______
d. ______
e. ______
f. ______
g. ______

a
b
c
d
e
f
g

Fig. 27-1.

______ 3. There are two types of wood-sill construction used over foundation walls. (T or F)

______ 4. Sills should be attached to the wall with 1/2" ______.

______ 5. There should be at least ______ bolts in each pair of sills.

______ 6. Joists are usually placed ______ inches on center.

______ 7. The header joist should be aligned with the outside edge of the ______.

______ 8. For openings in floor framing for such things as stairs, chimneys, or fireplaces, the framing members should be tripled. (T or F)

______ 9. For a bay window, floor ______ should extend beyond the foundation wall.

______ 10. Beams placed at an angle to joists are called ______.

(Continued on next page)

_______________ 11. Bridging stiffens the floor and distributes the load evenly on a joist. (T or F)

_______________ 12. Bridging can be either wood or metal. (T or F)

_______________ 13. Girder floor framing is different from conventional floor framing in the following ways *(one wrong)*: a. better for cold climates; b. can be built faster; c. members must be heavier; d. popular for houses with no basement.

_______________ 14. Subflooring is laid directly over the floor _______________.

_______________ 15. Plywood makes an excellent subflooring. (T or F)

_______________ 16. Floor framing can be held together with construction adhesives. (T or F)

_______________ 17. Modern job sites commonly use _______________ or OSB for subflooring.

_______________ 18. The most common metal connector used in floor framing is the joist _______________.

_______________ 19. Floor trusses are made in a factory to job specifications. (T or F)

_______________ 20. The common depths of floor trusses are 14″ and _______________″.

_______________ 21. Floor trusses can be long enough to reach from one side of the house to the other. (T or F)

_______________ 22. A parallel-chord truss has _______________ basic parts.

_______________ 23. The top and bottom of a parallel-chord floor truss are called _______________.

_______________ 24. Another name for wood I-beams is _______________.

_______________ 25. I-beams are always built to specific lengths in a factory. (T or F)

_______________ 26. Wood I-beams should be stored on edge. (T or F)

_______________ 27. A block called a _______________ stiffener should be added to both sides of an I-beam.

____

_______________ 28. The top and bottom chords are connected by devices made of wood or galvanized _______________. Each of these connectors is called a _______________.

To be used with Unit 30

Pages 316-350

Name ______________________

Score: (43 possible) ______

Study Unit 28

Wall Framing

______ 1. The outside walls of a house *(one wrong)*: a. support the roof; b. act as framework for attaching exterior facing; c. are the insulation for the house; d. add design to the house's appearance.

______ 2. The exterior wall contains *(one wrong)*: a. interior and exterior coverings; b. windows and doors; c. insulation; d. studs; e. gussets.

______ 3. Wall-framing members are called studs. (T or F)

______ 4. Studs must be stiff, good at holding nails, and free from ______.

______ 5. Of the two general types of wall framing, ______ construction is used the most.

6. Match the items at the left with the descriptions at the right:

a. ______ a. studs
b. ______ b. sole plate
c. ______ c. top plate
d. ______ d. corner post
e. ______ e. trimmer studs
f. ______ f. header

1. forms both inside and outside corner
2. slender wood members placed vertically
3. laid horizontally to carry bottom end of studs
4. connecting link between wall and roof
5. support the header over an opening
6. horizontal member installed over an opening

______ 7. The members installed vertically over and under a window opening are called ______ studs.

______ 8. The allowance made for framing in doors and windows is called *(one right)*: a. header; b. top plate; c. rough opening; d. finish opening.

______ 9. The allowance made for a door or window is the distance between the ______ studs.

10. Name the parts of the wall framing shown in Fig. 28-1.

a. ______
b. ______
c. ______
d. ______
e. ______
f. ______

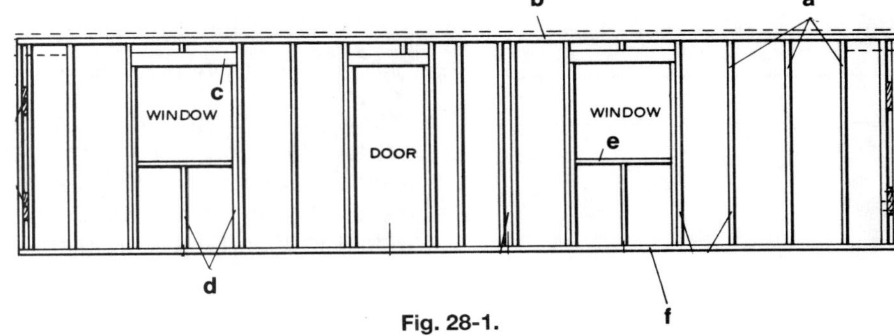

Fig. 28-1.

(Continued on next page)

a. ____________________
b. ____________________
c. ____________________
d. ____________________

11. The top plate does the following:
    a. Acts as a connection between the wall and ____________.
    b. Supports the lower ends of the ____________.
    c. Ties the studding together at the ____________.
    d. Forms a finish for the ____________.

a. ____________________
b. ____________________
c. ____________________

12. The location of studs should be determined from a common point because:
    a. Studs will then be directly over floor ____________.
    b. Ceiling joists and rafters will be directly over ____________.
    c. Members will be aligned from the rafter down to the ____________ wall of the building.

____________________ 13. A full-size layout that shows such things as floor level, ceiling height, and door and window elevations is called a ____________ pole.

____________________ 14. Corner posts should consist of more than one stud. (T or F)

____________________ 15. When an opening must be cut in the studs for a door or window, a ____________ is installed to support the lower ends of the top studs.

____________________ 16. For a window, the tops of the cut bottom studs are supported by a rough ____________.

____________________ 17. When building the exterior walls, the shorter walls are usually erected first. (T or F)

____________________ 18. In one method of wall framing, an entire wall can be assembled lying flat on the subfloor and then lifted into place. (T or F)

____________________ 19. To be sure the wall framing is straight, all exterior and intersecting ____________ should first be plumbed.

____________________ 20. The straightening is accomplished with either a ____________ or a plumb bob.

____________________ 21. Temporary ____________ is nailed in place to keep the walls straight.

____________________ 22. A nonbearing wall is one that supports ceiling joists. (T or F)

____________________ 23. Walls that divide the inside space of a house are called ____________.

____________________ 24. Special framing is required for openings for *(one wrong)*: a. windows; b. pipes; c. bathtubs; d. heating vents.

____________________ 25. Special framing requires extra material for strength. (T or F)

____________________ 26. The bulkhead above a cabinet is called a ____________.

____________________ 27. Cabinets can be built on the job or ____________.

____________________ 28. In platform framing *(one wrong)*: a. the second floor of a multilevel house has a platform on which the carpenter can work; b. ceiling joists for the first floor are the floor joists of the second floor; c. wall studs go from the sill plate to the roof rafters; d. ceiling joists of the first floor must be wider to support the second story.

To be used with Unit 31

Pages 351-357

Name ______________________

Score: (33 possible) ______

Study Unit 29

Structural Wall Sheathing

______ 1. The inner layer of the outside wall is called ______.

______ 2. Siding is a structural element of the wall. (T or F)

______ 3. Sheathing is part of the wall framing. (T or F)

______ 4. Insulating sheathing requires diagonal corner bracing. (T or F)

______ 5. Structural sheathing is a structural element of the house. (T or F)

______ 6. Three common kinds of structural sheathing are wood, plywood, and ______.

______ 7. Structural sheathing requires corner bracing. (T or F)

______ 8. Structural sheathing that covers large areas and adds great strength and rigidity is ______.

______ 9. Plywood sheathed walls are ______ as strong as walls sheathed with diagonal boards.

______ 10. Plywood sheathing must be applied horizontally. (T or F)

______ 11. Plywood sheathing comes in sheets ______′ by 8′ or longer in length.

______ 12. The following are types of edges on wood sheathing *(one wrong)*: a. shiplap; b. square; c. angle; d. dressed and matched.

______ 13. Wood sheathing must have a ______ edge for butt joinery.

______ 14. Wood sheathing that is dressed and matched has tongue-and-groove joints. (T or F)

______ 15. Oriented-strand board can be used for structural sheathing. (T or F)

______ 16. OSB is generally used as a direct replacement ______.

______ 17. Gypsum sheathing consists of two layers of lightweight ______ with gypsum filler in between.

______ 18. Gypsum sheathing comes in ______′ × 8′ sheets.

______ 19. Gypsum sheathing should be applied vertically. (T or F)

20. Sheathing can be applied at these different times:

a. ______ a. When the ______ has been framed or after the roof frame has been cornered.

b. ______ b. When the wall is completely framed and squared and is lying on the ______.

c. ______ c. When the ______ frames have been erected, plumbed, and braced.

(Continued on next page)

______________________ 21. Carpenters prefer to add sheathing as soon as possible because *(one wrong)*: a. scaffolding is then available for framing the roof; b. the structure is then more rigid; c. the building is warmer to work in; d. there is a more solid structure for ceiling and roof.

______________________ 22. Wood sheathing may be applied either horizontally or ______________.

______________________ 23. Gypsum sheathing is nailed with 1 3/4″ or 2″ ______________ roofing nails.

______________________ 24. Plywood sheathing should be a minimum of ______________ thick.

______________________ 25. Plywood is usually applied ______________.

______________________ 26. Building paper is applied between the sheathing and the siding. (T or F)

______________________ 27. Building paper should be water-resistant but not ______________-resistant.

______________________ 28. Building paper should be applied ______________.

______________________ 29. House wraps are made of paper. (T or F)

______________________ 30. House wraps allow water vapor to pass through. (T or F)

______________________ 31. House wrap is easy to rip. (T or F)

To be used with Unit 32 Name ______________________

Pages 358-364 Score: (23 possible) ______

Study Unit 30
Ceiling Framing

______________ 1. Fig. 30-1 is an example of ____________ roof construction, one of the basic methods of roof framing for platform-frame construction.

a. ______________ 2. The names of the members in Fig. 30-1 are:

b. ______________

c. ______________

d. ______________

e. ______________

Fig. 30-1.

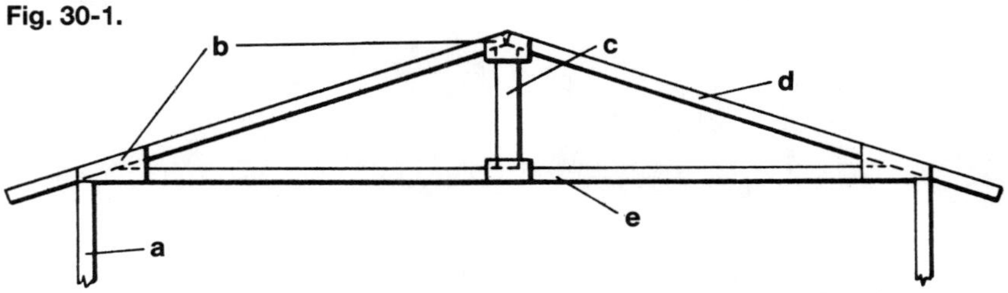

______________ 3. Besides the type shown in Fig. 30-1, another method of roof framing is called ____________ roof construction.

______________ 4. The parallel beams that support ceiling loads are called ____________.

______________ 5. The size of the ceiling joists depends upon *(one wrong)*: a. kind and grade of wood; b. distance they must span; c. load they must carry; d. local supplies available.

______________ 6. Ceiling joists form the ceiling of the house and support the ceiling finish. (T or F)

______________ 7. Except for the first two ceiling joists, the spacing should be 16 or ____________ inches on center.

______________ 8. When joining ceiling joists over the partition plate, a plywood or metal joist ____________ may be nailed to both sides of the joists.

______________ 9. Ceiling joists may also be joined so that they ____________ each other.

______________ 10. When attaching the ceiling joist to the exterior wall plate, it is important to keep the end of the joist even with the ____________ edge of the plate.

______________ 11. It is possible for a house to have a large open interior when ____________ are used.

______________ 12. When ceiling joists and rafters are used, the interior ends of the ceiling joists must be supported by a ____________.

______________ 13. Ceiling joists with a long span can be given additional support by the construction of a ____________.

(Continued on next page)

_______________ 14. The board feet of lumber needed for the joists of a house that contains 800 sq. feet of ceiling (joists are 2′ × 8′ and 24′ on center) is _______________ board feet.

_______________ 15. Prefabricated assemblies placed on framing are called _______________.

_______________ 16. Ceiling joists can serve as floor joists for an attic. (T or F)

_______________ 17. The maximum allowable length for ceiling joists that are 2″ × 4″ and spaced 12 inches apart would be 11′ 6″ of Group 1 lumber. (T or F)

_______________ 18. The number of board feet in lumber that is 2″ × 4″ and 12′ long is _______________ board feet.

_______________ 19. The number of board feet in a 2″ × 6″ ceiling joist that is 16′ long is _______________ board feet.

To be used with Unit 33

Pages 365-371

Name ______________________________

Score: (35 possible) ______

Study Unit 31

Roof Framing

a. ______________________
b. ______________________
c. ______________________
d. ______________________
e. ______________________
f. ______________________
g. ______________________
h. ______________________
i. ______________________
j. ______________________

1. Identify the roof styles shown in Fig. 31-1.

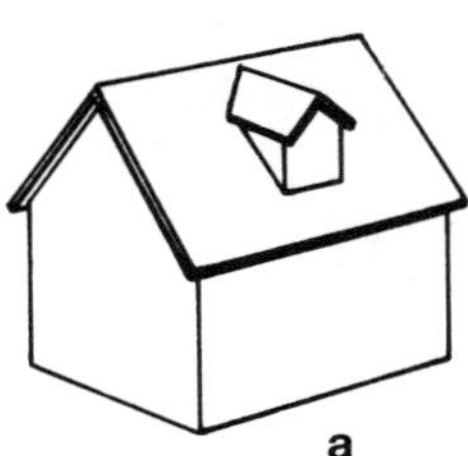
a

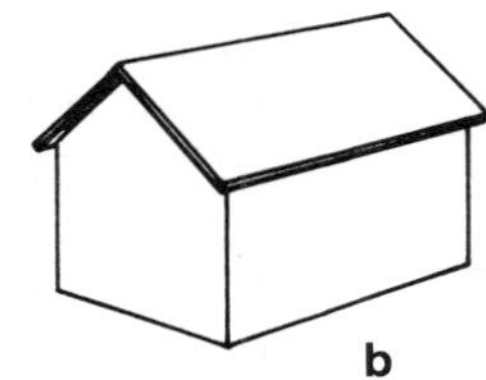
b

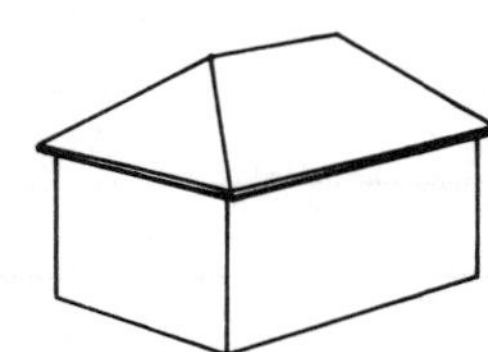
c

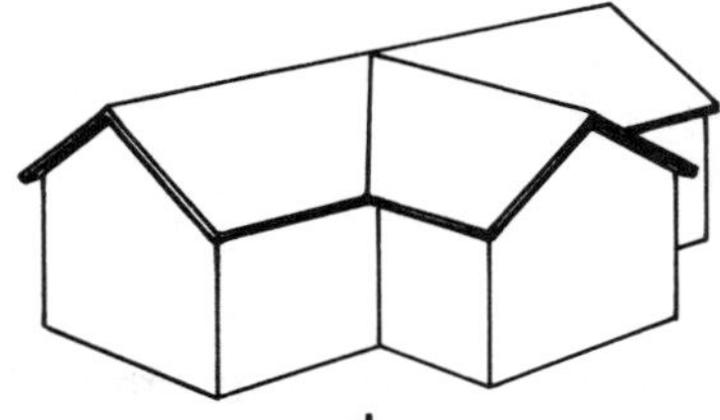
d

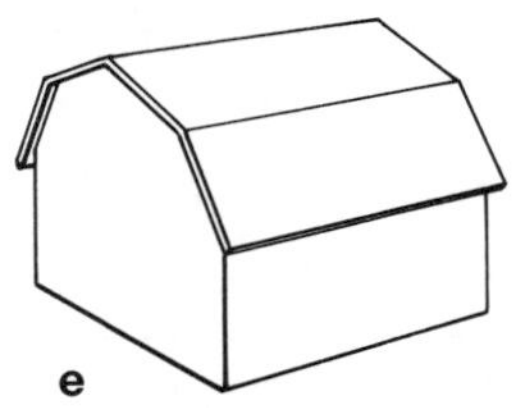
e

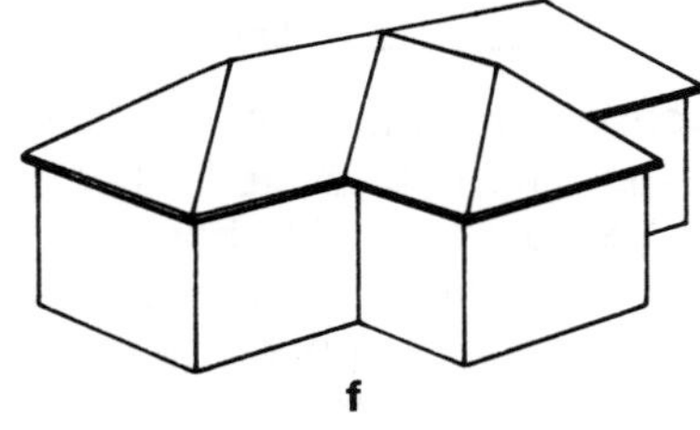
f

g

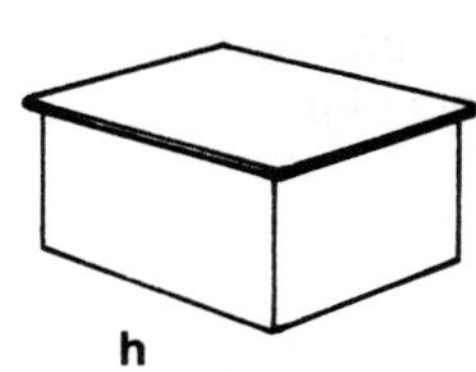
h

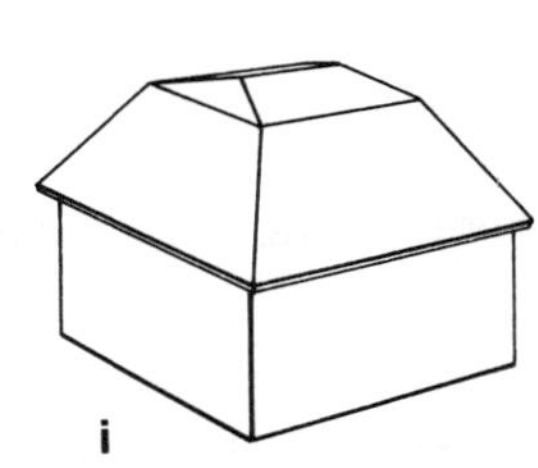
i

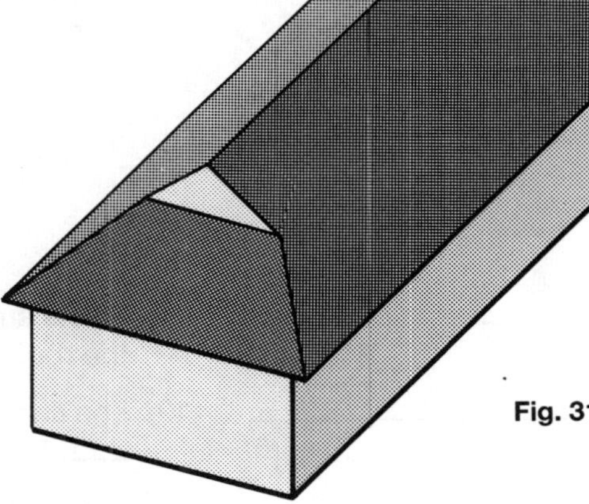
j

Fig. 31-1.

a. ______________________
b. ______________________
c. ______________________
d. ______________________

2. A roof should:
   a. Protect the house in all kinds of ____________.
   b. Require a minimum amount of ____________.
   c. Be strong enough to bear the load of snow and ____________.
   d. Be anchored to ____________ walls.

(Continued on next page)

_______________ 3. Architectural design bears little relationship to roof style. (T or F)

_______________ 4. A horizontal piece that connects the upper ends of the rafters is called the _______________ board.

_______________ 5. The overhanging part of a rafter is called the _______________.

_______________ 6. Another name for the incline of a roof is _______________.

7. Match the roof framing terms at the left with the descriptions at the right:

a. ______	a. pitch	1. one-half the distance of the span
b. ______	b. span	2. the rise in inches and the unit of run
c. ______	c. unit of run	3. an imaginary line that runs lengthwise from the outside wall to the ridge
d. ______	d. cut of a roof	4. the steepness of the roof
e. ______	e. measuring line	5. equal to 1′, or 12″
f. ______	f. unit rise	6. vertical distance from the top of the double plate to the upper end of the measuring line
g. ______	g. run	7. number of inches the roof rises per foot of run
h. ______	h. rise	8. distance between the outside edge of the double plates

_______________ 8. A line that runs vertical to a rafter is called a _______________ line.

_______________ 9. A line that is level with a rafter is called a _______________ line.

10. Match the rafters at the left with the descriptions at the right:

a. ______	a. jack	1. extend from the plate to the ridge board at 90° to both
b. ______	b. common	2. never extend full distance from the plate to the ridge board
c. ______	c. valley	3. extend diagonally from the corners formed by the plate to the ridge board
d. ______	d. hip	4. extend diagonally from the plates to the ridge board along lines where two roofs intersect

_______________ 11. Joists support the floor of a house the same as _______________ support the roof.

_______________ 12. The carpenter should have a roof frame plan to know what kind of _______________ are needed to frame the roof.

_______________ 13. The exact number of each kind of rafter can be determined from the scale drawing. (T or F)

To be used with Unit 34

Pages 372-380

Name ____________________

Score: (25 possible) ______

Study Unit 32

Conventional Roof Framing with Common Rafters

______ 1. Pitched roofs are built by four methods. (T or F)

______ 2. Figure 32-1 shows the ______ method of roof framing.

______ 3. In the conventional method of roof framing, joists and rafters are put up a piece at a time. (T or F)

______ 4. In ______ roof construction, the roof framing members are usually prefabricated.

Fig. 32-1.

______ 5. Some advantages of the joist and rafter method are *(one wrong)*: a. insulation can be easily installed between joists; b. roof load is carried on the walls; c. it can be done quickly; d. common materials are utilized for sheathing and finish.

______ 6. Some disadvantages of joist and rafter construction are *(one wrong)*: a. it takes a long time; b. the building is exposed to the weather longer; c. builder must utilize scaffolding; d. it requires special sheathing.

______ 7. In joist and rafter roof framing, the rafters are put in place after the ______ have been fastened in place.

______ 8. If it weren't for the ______, the rafters would spread and push out the exterior walls.

9. Complete the following statements on laying out a common rafter:

a. ______ a. The top of the rafter rests against the ______ board.

b. ______ b. The cut at the top of the rafter is called a top or ______ cut.

c. ______ c. The bottom of the rafter rests on the ______.

d. ______ d. The cut on the bottom of the rafter is called the level or ______ cut.

e. ______ e. Draw the ______ line along the edge of the tongue.

10. Three ways of figuring the length of a common rafter are:

a. ______ a. By stepping off the length with a ______ square.

b. ______ b. By using the ______ theorem.

c. ______ c. By applying the unit length obtained from the ______ table on the framing square.

(Continued on next page)

__________ 11. The theoretical length of a rafter and the actual length are the same. (T or F)

__________ 12. In a house that is 28′ wide with a 1/2 pitch roof, the theoretical length of a common rafter is *(one right)*: a. 124″; b. 188″; c. 215″; d. 230″.

13. The step-off method for finding the theoretical rafter length is done as follows:

a. __________ a. Place the square on the rafter with the tongue on the __________ cut.

b. __________ b. Step off the cut of the roof on the rafter stock as many times as there are __________ in the total run.

__________ 14. The length of a rafter as shown on the rafter table must be reduced by half the thickness of the ridge board. This is called a ridge __________.

__________ 15. When a rafter has an overhang, it has a notch called a __________.

__________ 16. The bird's-mouth and the plumb cut at the ridge can be eliminated from a rafter by using a __________ brace.

__________ 17. Two rafters should be cut and used to see how the heel cut and top cut fit before cutting remaining rafters. (T or F)

__________ 18. Rafters are usually leaned against the building with the ridge cut down. (T or F)

To be used with Unit 35

Pages 381-394

Name ______________________________

Score: (34 possible) ______

Study Unit 33

Hip and Valley Rafters

______________________ 1. The roof member that forms a raised area is called a ______________ rafter.

______________________ 2. The roof member that forms a depression in the roof is called a ______________ rafter.

______________________ 3. Rafters that form a raised area or depression in the roof should be ______________ inches deeper than common rafters.

4. Give the names of the parts drawn in the hip roof in Fig. 33-1.

a. ______________________

b. ______________________

c. ______________________

d. ______________________

e. ______________________

f. ______________________

g. ______________________

h. ______________________

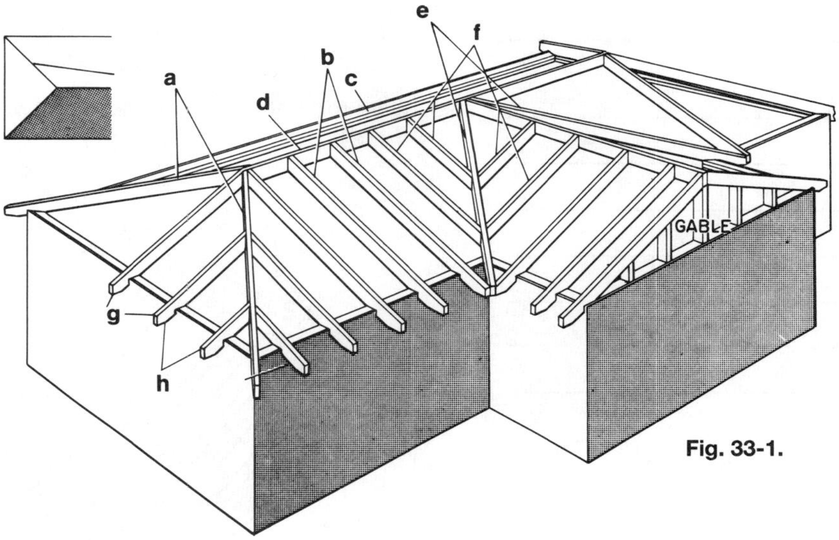

Fig. 33-1.

______________________ 5. The unit run of a hip rafter is the hypotenuse of a ______________ triangle with the shorter sides each equal to the unit run of a common rafter.

______________________ 6. The unit length of a hip rafter may be found on the third line of the rafter table on the steel square. (T or F)

______________________ 7. The plumb and level lines on a hip or valley rafter are also referred to as the top and ______________ cuts.

______________________ 8. If a hip rafter is framed against a ridge board, the shortening allowance is ______________ the 45° thickness of the ridge piece.

______________________ 9. The hip rafter joins the ridge piece or the ridge end of the common rafter at an ______________.

______________________ 10. The top line of the rafter table on the framing square is headed "Side Cut Hip or Valley Use." (T or F)

______________________ 11. The hip or valley rafter overhang is the same as the overhang for a common rafter on the same roof. (T or F)

(Continued on next page)

______________________ 12. The bird's-mouth is formed by two cuts, namely, the heel cut and the ______________ cut.

______________________ 13. In adjusting the upper edges of a hip rafter so that it will not interfere with the application of sheathing, two methods can be used, namely, ______________ and dropping.

______________________ 14. Dropping means to deepen the bird's-mouth to bring the top edge of the hip rafter down to the upper ends of the jacks. (T or F)

______________________ 15. Most roofs that contain valley rafters are equal-pitch roofs. (T or F)

______________________ 16. In framing an equal-span roof addition, the span of the addition is the same as the span of the main roof. (T or F)

______________________ 17. An unequal-span roof addition can be framed with one long and one short ______________ rafter.

______________________ 18. There are two methods of framing an unequal-span roof addition. (T or F)

______________________ 19. When framing a dormer without side walls, the upper edges of the headers must be beveled to the cut of the main roof. (T or F)

20. Give the names of the parts shown for a dormer in Fig. 33-2.

a. ______________________

b. ______________________

c. ______________________

d. ______________________

e. ______________________

f. ______________________

g. ______________________

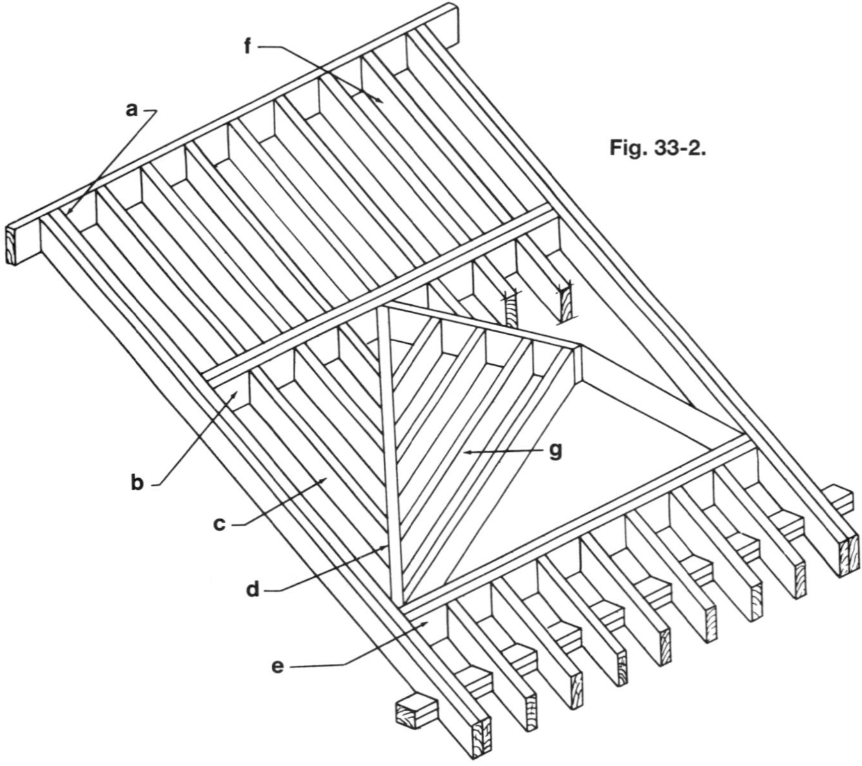

Fig. 33-2.

______________________ 21. A dormer may be constructed with side walls. (T or F)

To be used with Units 36 and 37 **Name** ____________________

Pages 395-419 **Score: (44 possible)** ______

Study Unit 34

Jack Rafters and Roof Framing

______ 1. A jack rafter is a ______ common rafter.

______ 2. The following are the common kinds of jack rafters *(one wrong)*: a. hip jack; b. common jack; c. valley jack; d. cripple jack.

______ 3. When framing an equal-pitch roof, the unit rise of a jack rafter is the same as the unit rise of a common rafter. (T or F)

______ 4. There are ______ types of cripple jack rafters.

______ 5. Jack rafters are usually spaced 12 inches apart. (T or F)

______ 6. The length of the shortest hip jack when the jacks are spaced 16 inches on center and the unit rise is 8 inches is ______ inches.

______ 7. The best way to figure the total length of valley jacks and cripple jacks is to lay out a framing diagram. (T or F)

______ 8. In Fig. 36-4 in the text the run of valley jack No. 1 is the same as the run of hip jack No. 7. (T or F)

______ 9. In Fig. 36-4 in the text the run of valley jack No. 2 is the same as the run of hip jack No. 7. (T or F)

______ 10. In Fig. 36-4 in the text the run of valley cripple No. 14 is ______ the run of valley cripple No. 13.

______ 11. A hip jack rafter has a shortening allowance at the upper end consisting of one-half the 45° thickness of the hip rafter. (T or F)

______ 12. The bird's-mouth and overhang of a jack rafter are laid out the same as on a common rafter. (T or F)

______ 13. Before a building is ready for roof framing, the following must be completed *(one wrong)*: a. all framing is complete; b. all framing is plumbed and squared; c. exterior siding is applied; d. the ceiling joists are in place.

______ 14. On a gable roof the theoretical length of the ridge piece is equal to the length of the building. (T or F)

______ 15. On a gable roof, if there is an overhang, the ridge board is shorter than the length of the building. (T or F)

______ 16. A hip roof may have an equal or an unequal-span addition. (T or F)

______ 17. The ridge board will be the same length both in equal-span and in unequal-span additions. (T or F)

______ 18. The length of the ridge piece on a dormer is the same whether or not it has side walls. (T or F)

______ 19. The layout of rafter spacing can be found by checking the building plans or the roof framing plans. (T or F)

(Continued on next page)

_______________ 20. For a gable roof the rafter locations are laid out on the ridge piece first. (T or F)

_______________ 21. The rafters for a gable roof should butt directly opposite each other on the ridge board. (T or F)

_______________ 22. There are two common methods of erecting the ridge board. (T or F)

_______________ 23. Roof framing should be done from a scaffold with planking not less than _______________ feet below the level of the main roof ridge board.

_______________ 24. On a hip roof the ridge board and the common rafters extending from the ridge ends to the side walls are erected first. (T or F)

_______________ 25. Common rafters on a hip roof must be plumbed. (T or F)

_______________ 26. Hip rafters on a hip roof are toenailed to the plate corners with _______________ nails, two on each side.

_______________ 27. All hip jacks should be nailed on one side first and then all the jacks on the opposite side. (T or F)

_______________ 28. Hip jacks are toenailed to hip rafters with 10d nails, using _______________ for each jack.

_______________ 29. Gable or double-pitch roof rafters are often reinforced by horizontal members called _______________ beams.

_______________ 30. The theoretical and actual lengths of the collar beams are identical. (T or F)

_______________ 31. A gable roof may be framed with or without an overhang. (T or F)

_______________ 32. A gable roof with two slopes is called a _______________ roof.

_______________ 33. An advantage of the gambrel roof is that it provides additional space for rooms in the attic. (T or F)

_______________ 34. A gambrel maximizes the roof area exposed to snow loads. (T or F)

_______________ 35. A shed roof is essentially one-half of a _______________ roof.

_______________ 36. A dormer may be framed into a shed roof. (T or F)

_______________ 37. A flat roof is always perfectly flat. (T or F)

_______________ 38. A common kind of construction used with flat or low-sloped roofs is the post-and-_______________ construction.

_______________ 39. It is possible to combine the ceiling and roof elements into one system on a flat or low-sloped roof. (T or F)

_______________ 40. Flat and low-pitched roofs usually require larger sized rafters than pitched roofs. (T or F)

_______________ 41. Roof openings are commonly required for the following *(one wrong)*: a. ventilator; b. overhang; c. chimney; d. skylight.

_______________ 42. A chimney saddle can be fabricated on the ground. (T or F)

_______________ 43. The number of rafters necessary for a building can be counted directly from a roof framing plan. (T or F)

_______________ 44. It is not possible to estimate the number of rafters necessary for a house. (T or F)

To be used with Unit 38 — Name ____________________

Pages 420-433 — Score: (34 possible) ______

Study Unit 35

Roof Trusses

1. A trussed rafter is the same as a simple truss. (T or F) ______
2. A simple truss is:
   a. Pieces put together to form a stiff framework of ______ shapes. a. ______
   b. Connected at the joists by ______. b. ______
   c. Able to hold loads over a long span without intermediate ______. c. ______
3. Gussets join together the parts of a truss by means of nails, screws, bolts, or ______. ______
4. Trusses usually come to the building site preassembled. (T or F) ______
5. The use of roof trusses adds considerably to material and labor costs. (T or F) ______
6. The use of trusses has the following advantages:
   a. No interior bearing ______ are needed. a. ______
   b. ______ can be placed anywhere in the home interior. b. ______
   c. Trusses can be put up very ______. c. ______
7. The types of trusses used for home construction include *(one wrong)*: a. scissors; b. Y-type; c. king-post; d. W-type. ______
8. Identify the trusses shown in Fig. 35-1.
   a. ______
   b. ______
   c. ______

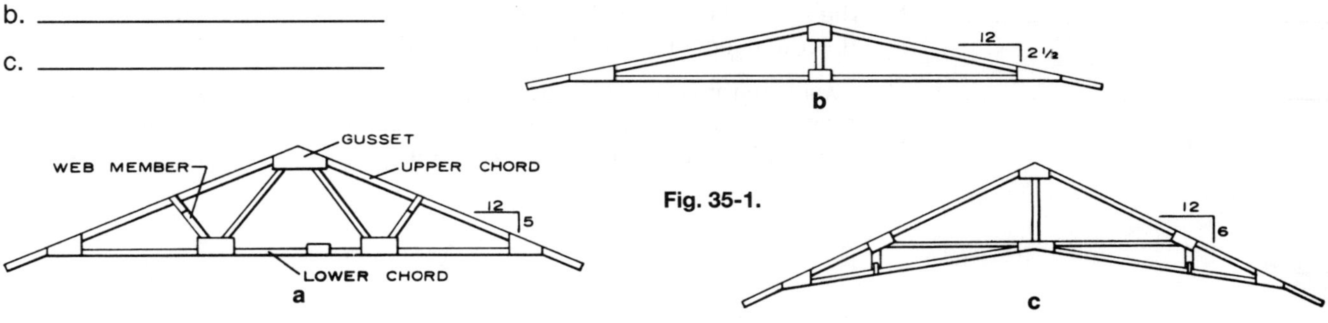

Fig. 35-1.

9. Match the items at the left to the descriptions at the right:

a. ______	a. scissors	1. more economical for shorter spans because it has fewer pieces
b. ______	b. king-post	2. used for "cathedral" ceiling
c. ______	c. W-type	3. has more pieces with less distance between connections

(Continued on next page)

10. The common spacing for roof trusses is ____________ inches.

11. The truss used the most is the ____________.

12. Factors most important in truss design are *(one wrong)*: a. kind of wood; b. snow and wind loads; c. slope of roof; d. weight of roof itself.

13. Gusset plates used on trussed rafters are made of metal or ____________.

14. The moisture content of a wood truss should not exceed ____________ percent.

15. Follow these instructions when assembling a truss on the job:

a. When mixing and using glue, follow the ____________ directions.

b. Do the gluing under controlled ____________ conditions.

c. Put the truss aside in an ____________, inverted position.

16. Nails or ____________ may be used to supply pressure until the glue is set when applying the gussets.

17. A truss can be put into position in the roof immediately after assembly. (T or F)

18. Building trusses on the job requires extensive equipment. (T or F)

19. When building a truss, don't forget to provide for a slight curvature, called ____________.

20. Large metal gussets used for commercial trusses are called truss ____________.

21. A truss can be fastened to the wall plates *(one wrong)*: a. by toenailing; b. by a metal bracket; c. with a framing anchor; d. with adhesive.

22. The best method for fastening a truss to the wall plate is *(one right)*: a. by toenailing; b. by a metal bracket; c. with a framing anchor; d. with adhesive.

23. The amount of material needed for a roof truss is determined the same as for conventional roof framing. (T or F)

24. It will take three workers about ____________ hours to install roof trusses on an average size home, with attached garage, containing about 2,000 square feet of ceiling area.

Answers:

10. ____________________
11. ____________________
12. ____________________
13. ____________________
14. ____________________
15. a ____________________
b. ____________________
c. ____________________
16. ____________________
17. ____________________
18. ____________________
19. ____________________
20. ____________________
21. ____________________
22. ____________________
23. ____________________
24. ____________________

To be used with Unit 39 Name ______________________

Pages 434-443 Score: (41 possible) ______

Study Unit 36
Roof Sheathing

1. Roof sheathing is considered a structural part of a building. (T or F) ______
2. Lumber roof sheathing and ______ are two good sheathing choices for homes with pitched roofs. ______
3. Lumber or ______ roof decking is sometimes used in homes with exposed ceilings. ______
4. Manufactured fiber roof decking can be adapted to ______ ceiling applications. ______
5. The hardwoods are the wood species used for roof sheathing boards. (T or F) ______
6. Roof sheathing boards to be covered with asphalt shingles do not have to be seasoned. (T or F) ______
7. Board roof sheathing is always laid horizontally. (T or F) ______
8. When roof sheathing boards are laid with no spaces between, it is called ______ sheathing. ______
9. Roof boards should give solid and continuous support for *(one wrong)*: a. roofs where snow and wind-driven conditions prevail; b. roofs made of asphalt shingles; c. roofs made of wood shakes; d. roofs made in cold climates. ______
10. Lumber roof sheathing is nailed to each rafter with two ______ nails. ______
11. Boards should be laid in such a way that they bear on at least two rafters. (T or F) ______
12. As a sheathing material, plywood:
    a. Provides a smooth, solid ______. a. ______
    b. Provides a minimum number of ______. b. ______
    c. Costs less by cutting ______ to a minimum. c. ______
    d. Offers flexibility in ______. d. ______
13. Plywood sheathing is very durable. (T or F) ______
14. One disadvantage of plywood sheathing is the considerable time it takes to install it. (T or F) ______
15. Face grain of plywood sheathing should run in the same direction as the rafters. (T or F) ______
16. If plywood roof sheathing is not the exterior type, it should not be exposed to the weather. (T or F) ______

(Continued on next page)

a. ______________________

b. ______________________

c. ______________________

d. ______________________

17. Match the plywood thickness at the left to the roofs for which they should be used at the right:

a. 3/8″	1. slate or tile and rafter spacing of 24″
b. 1/2″	2. asphalt shingles and rafter spacing of 24″
c. 5/16″	3. asphalt shingles and rafter spacing of 16″
d. 5/8″	4. slate or tile and rafter spacing of 16″

______________________ 18. To help the worker apply plywood roof sheathing, a roof ____________ may be constructed.

19. Roof decking has the following good points:

a. ______________________ a. Is an excellent ____________ for any roofing material.

b. ______________________ b. Makes a ready-to-finish interior ____________.

c. ______________________ c. Provides a solid, ____________ roof deck.

d. ______________________ d. Is strong enough for use as ____________ decking.

e. ______________________ e. Comes in many ____________, patterns, and sizes.

______________________ 20. ____________ grade decking is the best choice when appearance is important.

______________________ 21. The grade of decking to choose when appearance and strength are not of prime importance is ____________ grade.

______________________ 22. Decking can be purchased with tongue and groove edges. (T or F)

______________________ 23. Decking comes in nominal widths of 4 to ____________ inches.

______________________ 24. Check the manufacturer's recommendations to get the correct ____________ for the span for all types of decking.

______________________ 25. Decking should be installed *(one wrong)*: a. with all the end joints in line; b. with each plank bearing on at least one support; c. with a 2° angle cut in the butting pieces; d. in a controlled random laying pattern for best economy.

______________________ 26. Decking should be fastened with common nails which are ____________ as long as the thickness of the nominal plank.

______________________ 27. ____________ common nails have better holding power than bright common nails.

______________________ 28. Metal ____________ are recommended on decking end joints of 3″ and 4″ material for better alignment, appearance, and strength.

______________________ 29. Roof decking made of wood fiber combines the advantages of ____________ and insulation.

______________________ 30. Wood fiber roof decking *(one wrong)*: a. is weatherproof; b. comes coated in a wide variety of colors; c. is ideal for shingles on all types of buildings; d. is fairly expensive.

______________________ 31. For a roof with a chimney, the sheathing should clear the brick by ____________ inch.

To be used with Unit 40

Pages 444-470

Name ______________________

Score: (57 possible) ______

Study Unit 37

Roof Coverings

______ 1. Roofing materials for pitched roofs include *(one wrong)*: a. shingles; b. hot asphalt or tar; c. slate or tile; d. galvanized metal.

______ 2. If a roof is shingled, the shingles should always overlap. (T or F)

3. Match the items at the left to the descriptions at the right:

a. ______
b. ______
c. ______
d. ______
e. ______
f. ______
g. ______

a. sidelap or endlap
b. coverage
c. toplap
d. square
e. shingle butt
f. exposure
g. headlap

1. lower exposed edge of a shingle
2. amount of protection overlapping shingles provide
3. amount of roofing needed to cover 100 sq. ft.
4. shortest distance in inches two shingles overlap each other horizontally
5. shortest distance in inches between exposed edges of overlapping shingles
6. the width of a shingle minus its exposure
7. shortest distance in inches from lower edges of overlapping shingle to upper edge of unit in second course below

______ 4. Slope and pitch are the same. (T or F)

______ 5. Many roofing accessories are needed including *(one wrong)*: a. underlayment; b. roofing cements; c. roofing nails; d. fascia.

______ 6. The amount of overlap of shingles depends largely on the kind of shingle and the ______ of the roof.

______ 7. A roof that rises 4″ for each 12″ of run is designated as a ______ slope.

______ 8. A roof shingled with slate or asphalt should have a slope of at least *(one right)*: a. 2 in 12; b. 3 in 12; c. 4 in 12; d. 6 in 12.

______ 9. A built-up roof should slope no more than *(one right)*: a. 2 in 12; b. 3 in 12; c. 4 in 12; d. 6 in 12.

______ 10. All types of shingles require application of underlayment. (T or F)

______ 11. A good underlayment is *(one right)*: a. waterproof paper; b. coated felt; c. laminated paper; d. asphalt-saturated felt.

______ 12. It is important to choose an underlayment with a ______ vapor barrier.

(Continued on next page)

13. Underlayment serves the following purposes:

a. ______________ a. Keeps asphalt shingles from direct contact with ______________ sheathing, which could cause a chemical reaction.

b. ______________ b. Protects roof sheathing from absorbing ______________ before shingles are applied.

c. ______________ c. Prevents ______________ from driving below shingles onto roof sheathing.

______________ 14. Underlayment should be applied immediately after ______________.

______________ 15. A metal or other special material applied to ward off water seepage is called ______________.

16. Some areas that need material to prevent water seepage are:

a. ______________ a. At the intersection between roof and soil stack or ______________.

b. ______________ b. Around ______________.

c. ______________ c. At the point where a ______________ intersects a roof.

d. ______________ d. In the ______________ of a roof.

______________ 17. Homes in all climates require application of eaves flashing. (T or F)

______________ 18. The edges of a roof are protected from leaks by the installation of ______________ edges.

______________ 19. Roof shingles made of asphalt come in the following basic kinds *(one wrong)*: a. individual shingles of a very large size; b. individual shingles, either interlocking or staple-down; c. asphalt and wood combination; d. strip shingles.

______________ 20. If a roof is longer than 30′, asphalt shingles should be applied starting at the center. (T or F)

______________ 21. On a roof shorter than 30′, asphalt shingles can be applied from either ______________.

______________ 22. The pattern of laying asphalt strips can be varied with full or cut strips. (T or F)

______________ 23. Square-butt strip shingles are laid in variations, as follows *(one wrong)*: a. cutouts breaking joints on halves; b. cutouts breaking joints on thirds; c. cutouts breaking joints on fourths; d. random spacing.

______________ 24. When laying each course of shingles, the lower edges of the butts should be aligned with the top of the ______________ on the underlying course.

25. Techniques to follow for applying asphalt shingles to hips and ridges are:

a. ______________ a. Bend each shingle lengthwise down the ______________ with an equal amount on each side of the hip or ridge.

b. ______________ b. Begin at a ______________ or at one end of a ridge.

c. ______________ c. Use a ______________ inch exposure.

______________ 26. The closed method of application is always used for valleys. (T or F)

______________ 27. A roof with a 3-in-12 slope is considered low pitch. (T or F)

28. For a low-pitch roof, strip shingles should be applied with the following requirements:

a. ______________ a. Underlayment of ______________ thickness.

b. ______________ b. Flashing strip on the ______________ cemented in place.

c. ______________ c. ______________ applied to shingles at the factory.

______________ 29. Lock-down, also called ______________, shingles should be chosen for areas subject to high winds.

______________ 30. Hand-cut wood shingles are called ______________.

______________ 31. Hand-cut wood shingles come in lengths of *(one wrong)*: a. 12″; b. 18″; c. 24″; d. 32″.

______________ 32. Regular wood shingles come in lengths of *(one wrong)*: a. 16″; b. 18″; c. 24″; d. 32″.

______________ 33. Wood shingles should extend out over the eave or rake. (T or F)

______________ 34. Characteristics of roll roofing are *(one wrong)*: a. can be applied over old roofing; b. is less attractive in appearance than other roofing materials; c. is possibly less durable than asphalt shingles; d. is quite costly.

35. A typical built-up roof consists of:

a. ______________ a. Decking made of ______________.

b. ______________ b. Layers of roofer's ______________.

c. ______________ c. Each layer mopped down with asphalt or ______________.

d. ______________ d. Final coating of tar or ______________.

e. ______________ e. Topping of roofing ______________ or roll roofing.

______________ 36. An ideal roofing for garages and sheds is ______________ metal roofing.

______________ 37. Gutters must be made of metal. (T or F)

______________ 38. Metal gutters should be installed so they drain toward the downspouts at a slope of ______________ inch in 10 feet.

Name ______________________

Score: (24 possible) ______

Study Unit 38
Ventilation

______ 1. Trapped air and moisture inside a home are caused by *(one wrong)*: a. effective weatherstripping; b. laundry equipment on the first floor; c. roofs with wide overhang; d. use of sheet material for sheathing.

______ 2. Exterior walls must be protected against moisture from both inside and outside sources. (T or F)

______ 3. Outside walls are protected from moisture that comes from the inside of a home by a ______ barrier.

______ 4. The following materials provide suitable protection against vapor *(one wrong)*: a. sheet metal; b. porous paper; c. asphalt-laminated paper; d. foil laminates.

5. Match the areas of a house at the left to the moisture protection methods and techniques at the right:

a. ______
b. ______
c. ______
d. ______
e. ______

a. basement walls	1. soil should be covered with vapor-resistant material
b. crawl space	2. two coats of low-permeability paint should be applied
c. ceiling	3. asphalt coat should be applied to outside surfaces
d. basement floors	4. not needed except in very cold climates
e. interior of house with no vapor protection	5. layer of moisture protection should be placed underneath

______ 6. Condensation problems are greatly reduced by good ventilation. (T or F)

______ 7. Adequate use of lashing helps to assure against harmful moisture entering a home. (T or F)

______ 8. The size of the vent area in a home is relatively unimportant. (T or F)

9. Construction steps necessary for moisture protection are as follows:

a. ______ a. Exterior walls with no overhanging eaves should have ______.

b. ______ b. Poured concrete walls should have two coats of ______ or tar.

c. ______ c. The lot the house is built on should provide ______ away from the foundation walls.

d. ______ d. Basement floors and slab construction should rest on 4″ of ______ or stone covered with vapor barrier.

e. ______ e. In such places as under exterior windows, around chimneys, and between different exterior construction materials, ______ should be installed.

(Continued on next page)

10. Movement of air from inside to outside of a house can be provided by:

a. ____________ a. ____________ with screens in the gable ends of an attic.

b. ____________ b. A slot under eaves and a sheet-metal ____________ near the peak in a house with a hip roof.

11. The wood structure of a home can be protected from ground moisture as follows:

a. ____________ a. In a house with a crawl space, there should be at least ____________ inches of space between the ground and the wood framing.

b. ____________ b. Wood sills that rest on concrete or masonry walls should be at least ____________ inches above ground.

c. ____________ c. The outside siding, stairs, and trim should be at least ____________ inches above exposed earth.

d. ____________ d. Floor joists should be above the ground at least ____________ inches.

e. ____________ e. Floor framing for such things as porches and patios should be at least ____________ inches above the ground.

To be used with Unit 42

Pages 480-498

Name ______________________

Score: (26 possible) ______

Study Unit 39

Roof Trim

a. ______________________

b. ______________________

c. ______________________

1. Match the terms at the left with the descriptions at the right:
   a. cornice
   b. eave
   c. rake

   1. the edge at the end of a gable roof
   2. the rafter-end overhang of a roof
   3. finish on the outside just below the eave

______________________ 2. A gable roof has eaves on all four ends. (T or F)

______________________ 3. The three general types of cornice are close, open, and ______________.

______________________ 4. A strip that fits close under the eave is called a ______________.

5. Name the parts of the cornices in Fig. 39-1:

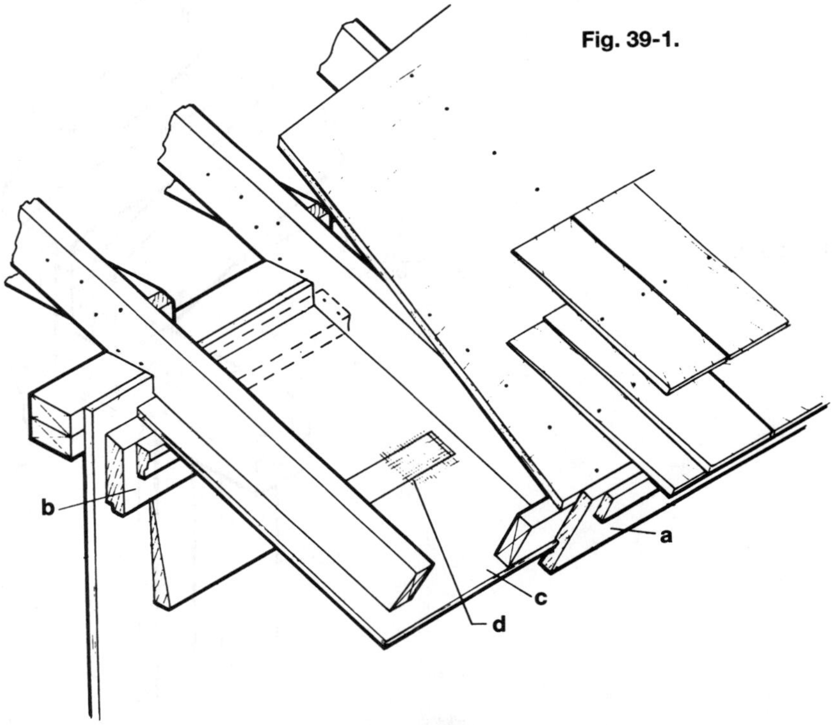

a. ______________________

b. ______________________

c. ______________________

d. ______________________

Fig. 39-1.

______________________ 6. A strip nailed to the tail plumb cuts of the rafters is called a ______________.

______________________ 7. In a ______________ cornice, the rafter overhang is entirely covered with roof covering, fascia, and bottom strip.

______________________ 8. The bottom strip is commonly referred to as a plancier, or ______________.

(Continued on next page)

a. ______________________

b. ______________________

c. ______________________

9. Identify the cornices shown in Fig. 39-2:

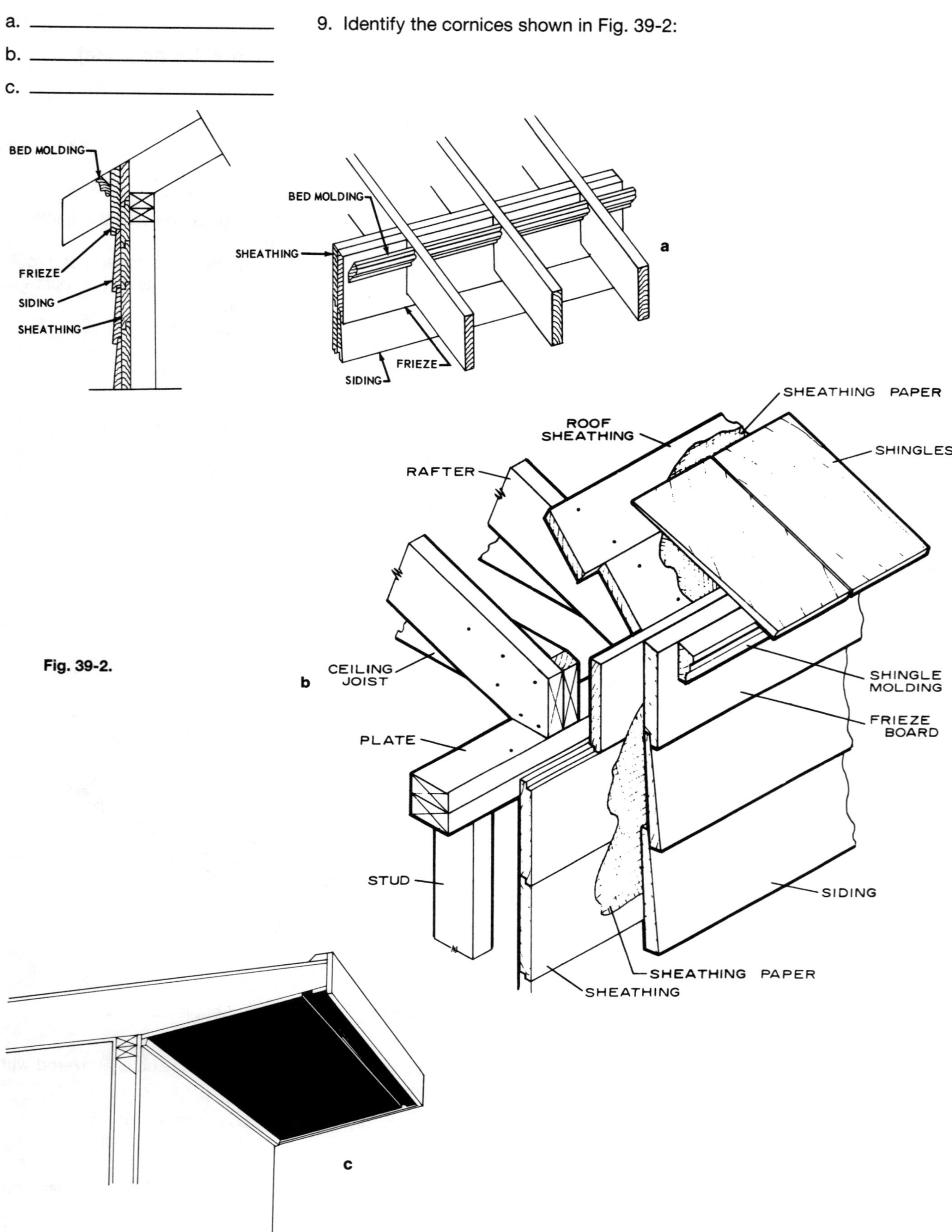

Fig. 39-2.

_______________ 10. The quality of workmanship for an open cornice is unimportant. (T or F)

_______________ 11. When constructing a box cornice, the rafter tails must be all in _______________.

_______________ 12. _______________ is a popular material that presents a smooth, attractive surface for constructing a box cornice.

_______________ 13. _______________ board is a noncombustible material developed for use in soffits.

_______________ 14. _______________ panels are often used for soffits on the undersides of eaves.

_______________ 15. A prefinished, low-maintenance, non-ferrous rust-free box cornice material is _______________.

_______________ 16. The ends of cornices on a gable roof must be finished. This finish is called the cornice _______________.

17. Match the cornice returns at the left to the descriptions at the right:

a. _______________

b. _______________

c. _______________

a. narrow cornice with box return
b. wide overhang at cornice and rake
c. narrow box cornice and close rake

1. used when there are wide overhangs at both sides and ends of the house
2. frieze board of the gable end joins the frieze board or fascia of the cornice
3. fascia board and shingle molding of the cornice are carried around the corner of the rake projection

To be used with Unit 43

Pages 499-530

Name ______________________

Score: (30 possible) ______

Study Unit 40

Windows and Skylights

a. ______________

b. ______________

c. ______________

d. ______________

e. ______________

f. ______________

1. Name the windows shown in Fig. 40-1.

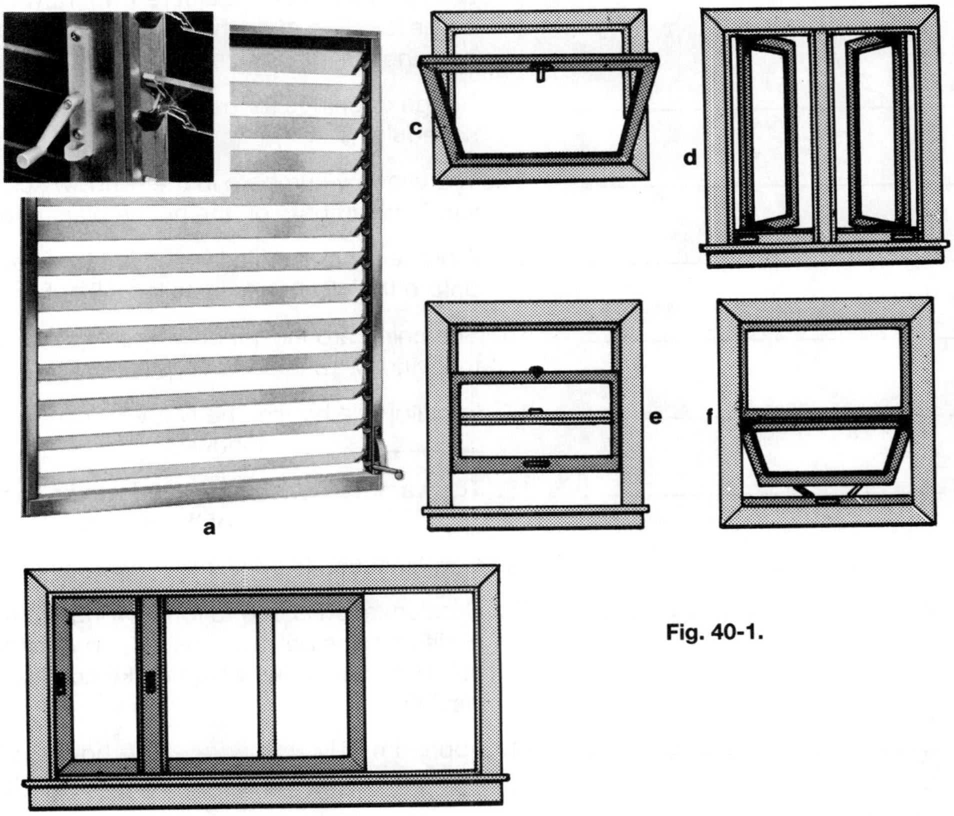

a

b

c

d

e

f

Fig. 40-1.

______________ 2. Windows are manufactured items brought to the building site fully assembled. (T or F)

______________ 3. Window frames are made of wood, aluminum, or ______________.

______________ 4. The window best known and most widely used is the ______________ window.

______________ 5. A window hinged at the sides that is designed to swing inward or outward by means of a crank is called the ______________ window.

______________ 6. An advantage of the ______________ window is better ventilation because the entire window opens.

______________ 7. A window designed solely for providing decoration and light is the ______________ window.

______________ 8. The ______________ window is hinged at the top so the bottom can swing out.

(Continued on next page)

______________________ 9. A window hinged at the bottom so the top can swing inward is called a ____________ window.

______________________ 10. Horizontal ____________ windows can be used effectively to create a wall of windows.

______________________ 11. ____________ are often built into a roof to provide ventilation and light.

______________________ 12. A window schedule contains the following information *(one wrong)*: a. descriptions of the different windows to be installed in the house; b. the location of each window; c. the cost of the windows; d. sash openings; e. glass sizes.

______________________ 13. Rough openings for each window are always given on the window schedule. (T or F)

______________________ 14. The window numbers in the window schedule correspond with the window numbers on the house plan. (T or F)

______________________ 15. A primer coat should be applied to wood window frame members before the window is installed. (T or F)

______________________ 16. Sheathing around window openings should always be applied flush with the window rough openings. (T or F)

______________________ 17. In building a house, the first windows to be installed are the ____________ windows.

______________________ 18. The frames of basement windows are an integral part of the ____________ wall of the house.

______________________ 19. Condensation is considered damaging to a home. (T or F)

______________________ 20. Condensation is due to *(one wrong)*: a. high humidity in the home; b. differences between inside and outside temperatures; c. improved building techniques which make houses more tightly built; d. poor insulation.

______________________ 21. Approximately ____________ hour(s) of labor will be needed for a window that contains 10 square feet or less of glass area.

______________________ 22. Skylights are installed to provide ____________ and light.

______________________ 23. Glazing of skylights can be either glass or ____________.

______________________ 24. It is easier to install a skylight in an existing roof than in a new roof. (T or F)

______________________ 25. All skylights can be opened for ventilation. (T or F)

To be used with Unit 44

Pages 531-554

Name ______________________

Score: (38 possible) ______

Study Unit 41

Exterior Doors and Frames

a. ______________________
b. ______________________
c. ______________________
d. ______________________
e. ______________________

1. Identify the doors shown in Fig. 41-1.

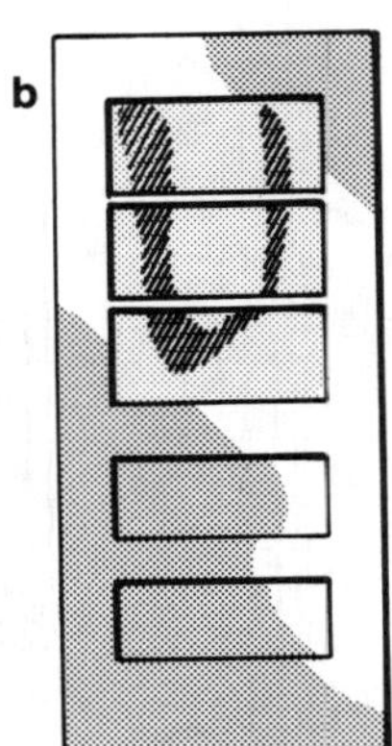

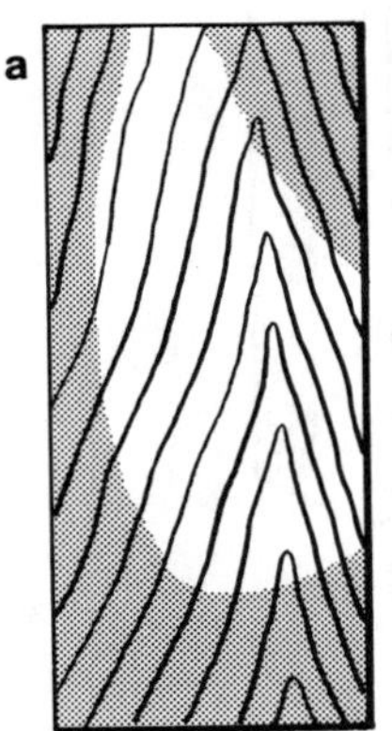

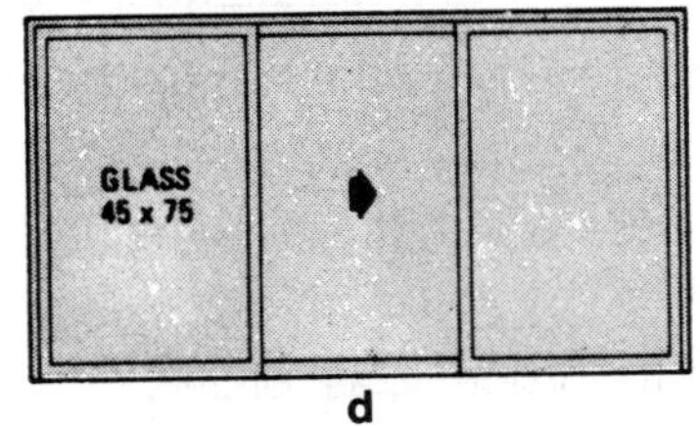

Fig. 41-1.

____

2. Another name for a French door is a ______________ door.

a. ______________________
b. ______________________
c. ______________________
d. ______________________
e. ______________________

3. Match the doors at the left to the descriptions at the right:
   - a. French
   - b. flush
   - c. panel
   - d. sliding
   - e. combination

   1. facing applied to a core
   2. panels including inserts of screen or storm
   3. set to ride in a track
   4. made of many panes
   5. made of thin members which fit between the rails and stiles

____

4. Exterior door thickness is *(one right)*: a. 1 1/2″; b. 1″; c. 1 3/4″; d. 2″.

____

5. Door jambs are always made of metal. (T or F)

(Continued on next page)

a. ______________________

b. ______________________

c. ______________________

d. ______________________

e. ______________________

f. ______________________

6. Name the parts of the doorframe shown in Fig. 41-2.

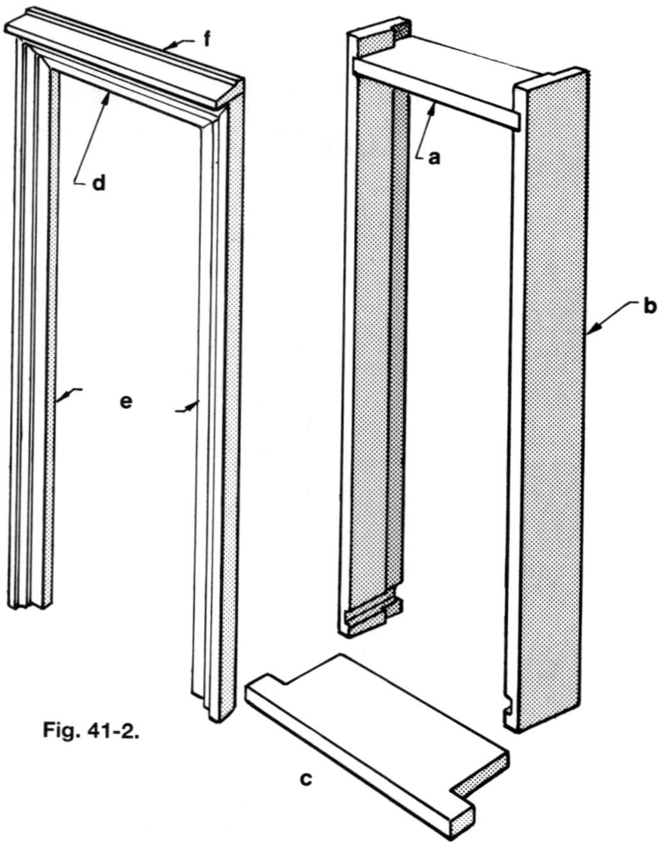

f

d

a

b

e

c

Fig. 41-2.

______________________ 7. The front or main entrance should be at least ____________ feet wide.

______________________ 8. The rough opening for the doorframe should measure the same size as the frame. (T or F)

______________________ 9. The doorframe should fit into the rough opening with the sill 2″ higher than the surface of the finished floor. (T or F)

10. Doors are expensive and should be cared for properly, as follows:

a. ______________________ a. Store under ____________ away from moisture.

b. ______________________ b. ____________ the top and bottom edges.

c. ______________________ c. Move a door by ____________ it.

d. ______________________ d. Wear ____________ when handling doors.

e. ______________________ e. Condition to the average ____________ content of the locality.

f. ______________________ f. Store on ____________ on a level surface to avoid warp or bow.

______________________ 11. Doors should be finish-painted at the same time as the woodwork. (T or F)

______________________ 12. A door should be fitted into the frame with a little leeway allowed for clearance. (T or F)

______________________ 13. The edge of the door on which the ____________ will go should be beveled.

______________________ 14. A device used for holding a door upright for planing and hardware installation is called a door ______________.

______________________ 15. The number of hinges usually installed on an exterior door is ______________.

16. Complete the following descriptions of garage doors:

a. ______________________ a. The door that opens outward and is held by door holders is the ______________ door.

b. ______________________ b. The door hung from a track above is called a sliding ______________ door.

c. ______________________ c. The ______________ door pivots on a track in the ceiling and has rollers at the center and top of the door.

d. ______________________ d. A door with rollers at each section fitted into a track at the side of the door and ceiling is the ______________ overhead door.

______________________ 17. The garage door best protected from weather is the ______________ door.

To be used with Unit 45
Pages 555-594

Name ______________________

Score: (62 possible) ______

Study Unit 42

Exterior Walls: Siding and Brick Veneer

1. The variety of exterior wall coverings available is due to:
   a. ______ a. The many different kinds of ______ used.
   b. ______ b. The many different shapes and ______.
   c. ______ c. The surface ______.

______ 2. Siding made of wood is about the most common of all exterior wall coverings. (T or F)

______ 3. Wood siding should *(one wrong)*: a. be easy to work; b. be made of hardwood; c. hold paint well; d. be free from warp.

______ 4. The most common types of wood siding are *(one wrong)*: a. shakes; b. board; c. bevel; d. drop.

______ 5. Wood siding that is nailed directly to the studs and that acts as both wall covering and sheathing is called ______ siding.

______ 6. Wood siding should be treated with water repellent before installation. (T or F)

______ 7. Bevel siding is always applied horizontally. (T or F)

______ 8. The amount of overlap of rows of bevel siding is the same for all widths. (T or F)

______ 9. A simple device that can be made to insure accuracy and increase efficiency in installing vertical siding is called a story pole. (T or F)

______ 10. When marking pieces of siding which will fit into spaces such as between two windows, a siding gauge called a ______ is very useful.

11. Wood siding nails should:
   a. ______ a. Be strong enough so ______ is not necessary.
   b. ______ b. Not discolor or ______ the siding.
   c. ______ c. Be easy to ______ into the siding.
   d. ______ d. Not cause ______, even at the edge.
   e. ______ e. Not pop out after being installed ______ with the siding.
   f. ______ f. Be ______ resistant.

12. The best wood siding nails are:
   a. ______ a. High tensile strength ______ nails.
   b. ______ b. ______ nails.

(Continued on next page)

a. ____________ b. ____________ c. ____________ d. ____________ e. ____________ f. ____________	13. Moisture will create problems with wood siding unless:
a. ____________	a. Joints in siding are sealed with ____________.
b. ____________	b. Siding pieces are cut carefully to insure proper ____________.
c. ____________	c. There is enough ____________ between courses.
d. ____________	d. Gutter joints and ____________ are well fitted.
e. ____________	e. ____________ is installed in places where horizontal surfaces meet siding.
f. ____________	f. The lowest edge of the siding is at least ____________ inches above the ground.
____________	14. The plywood best suited for siding is called medium-density ____________ plywood.
	15. Plywood makes a good siding because:
a. ____________	a. It is more ____________ and stronger than lumber sheathing.
b. ____________	b. It can be applied vertically or ____________.
c. ____________	c. It cuts the ____________ required for installation.
____________	16. Hardboard siding can be described as follows *(one wrong)*: a. tough; b. does not split or splinter; c. does not hold a finish as long as lumber; d. resistant to dents.
____________	17. Another siding material that is applied the same as hardboard is ____________ board.
____________	18. Sheathing is required when hardboard is installed. (T or F)
____________	19. Wood shingles can be purchased *(one wrong)*: a. in several grades; b. in standard widths; c. in several lengths; d. either finished or untreated.
____________	20. Wood shingles are applied to exterior walls by single- or double-coursing. (T or F)
____________	21. When wood shingles are applied over ____________ or fiberboard, horizontal nailing strips must first be nailed to the studs.
____________	22. It is recommended that wood shingles be no wider than ____________ inches.
____________	23. Wood shakes are made by ____________ different methods.
____________	24. ____________ shakes are made mostly by hand.
____________	25. Wood shingles and shakes can be allowed to weather. (T or F)
	26. Siding and shingles made of heat-resistant cement material must be applied:
a. ____________	a. Over ____________ sheathing.
b. ____________	b. Over ____________ paper that has been applied over the sheathing.
c. ____________	c. In the same way as horizontal ____________ siding.
____________	27. Stucco is a good finish for exterior walls. (T or F)

_______________ 28. Stucco should not be used for a two-story house unless the house was built by the _______ framing method.

29. When stucco is used as an exterior finish on a house:

a. _______________ a. The first coat should be forced through the _______.

b. _______________ b. It should be applied in _______ coats.

c. _______________ c. The temperature should be above _______ degrees F.

_______________ 30. A house covered with stone or brick veneer requires no sheathing. (T or F)

_______________ 31. Metal and plastic sidings have the following advantages *(one wrong)*: a. come with a baked-on finish; b. require little care and maintenance; c. can be worked with standard tools; d. require no sheathing.

_______________ 32. In applying metal or plastic siding, nails should be driven very tightly. (T or F)

33. Metal or plastic siding should be applied:

a. _______________ a. With an allowance for expansion and _______.

b. _______________ b. With overlaps in the siding away from areas of _______.

_______________ 34. Brick veneer is relatively low in initial cost. (T or F)

_______________ 35. The following are common mason's tools *(one wrong)*: a. trowel; b. level; c. hammer; d. shears.

_______________ 36. Common types of bricks include *(one wrong)*: a. building; b. facing; c. back; d. fire.

_______________ 37. All brick can be classified as either modular or nonmodular. (T or F)

_______________ 38. Type-O mortar is recommended for nonload-bearing walls. (T or F)

_______________ 39. The standard joint thicknesses are *(one wrong)*: a. 1/8″; b. 1/4″; c. 3/8″; d. 1/2″.

_______________ 40. A brick veneer wall must be supported on foundation. (T or F)

_______________ 41. Weep holes are placed in veneer walls to allow hot air to escape. (T or F)

_______________ 42. When new mortar is placed in an existing wall, it is called _______.

To be used with Unit 46

Pages 596-609

Name ______________________

Score: (38 possible) ______

Study Unit 43

Thermal Insulation, Radiant Barriers, and Vapor Barriers

______ 1. Insulation used as part of the construction of a home *(one wrong)*: a. keeps the home warmer in winter; b. keeps the home cooler in summer; c. increases the amount of duct work needed; d. reduces the size furnace required.

______ 2. A material having insulation properties controls energy in the form of heat, sound, and ______.

3. Match the insulation materials at the left to the descriptions at the right:

Answer	Materials	Descriptions
a. ______	a. reflective	1. applied by spraying or in board form
b. ______	b. corrugated paper	2. retards heat by means of radiation
c. ______	c. flexible	3. vegetable product made into sheets, squares, or boards
d. ______	d. rigid	4. blanket insulation made by spraying multiple layers of paper
e. ______	e. loose fill	5. made in the form of blankets or batts
f. ______	f. urethane or polystyrene	6. inorganic fibrous material blown against clean surfaces
g. ______	g. sprayed or foamed	7. poured, packed, or blown into such areas of home as attic and sidewalls

______ 4. Most flexible insulation is manufactured with a paper or sheet material covering. (T or F)

______ 5. Reflective insulation is better for cold climates than for warm. (T or F)

______ 6. Such materials as wallboard, roof decking, and building boards are considered examples of ______ insulation.

______ 7. A unit used to measure the transmission of ______ is called a Btu.

______ 8. A Btu is the amount of heat which will raise the temperature of one pound of water two degrees Fahrenheit. (T or F)

______ 9. Heating engineers and thermal experts have exact knowledge concerning the insulating qualities of recommended materials. (T or F)

______ 10. Heat loss from a house during cold weather can be reduced by insulating all walls, ceilings, roofs, and floors that separate heated from ______ spaces.

______ 11. Insulation for a house with an unheated crawl space should be placed between the floor joists or around the wall ______.

(Continued on next page)

__________ 12. A house with an unheated attic should have insulation installed around the stairway and in the first-floor __________.

__________ 13. A wall next to an unheated garage or __________ should be insulated.

__________ 14. A house with air conditioning has the same insulation requirements as a house built in a cold climate. (T or F)

__________ 15. Flexible insulation should be placed between framing members so that the tabs lap the edge of the studs and also the top and bottom __________.

__________ 16. The fastening device usually used for installing flexible insulation is a hand __________.

__________ 17. Reflective insulation should be installed tightly with no allowance for air space. (T or F)

__________ 18. Radiant barriers are not effective in hot climates. (T or F)

__________ 19. Radiant barriers must have a reflective material on both sides of the substrate. (T or F)

__________ 20. One-sided radiant barriers must have the reflective surface facing the open air space in the attic. (T or F)

__________ 21. The simplest method of installing a radiant barrier is on the attic floor. (T or F)

__________ 22. All material emits energy by __________ radiation.

__________ 23. The number 1 indicates the highest level of reflectivity. (T or F)

__________ 24. A radiant barrier can be attached near the roof of a house. (T or F)

__________ 25. Condensation of moisture can cause a problem when a radiant barrier is installed on the floor. (T or F)

__________ 26. The fire rating of radiant barriers must be rated class __________ by the National Fire Protection Association.

__________ 27. Vapor barrier should be put in place before application of loose fill insulation. (T or F)

28. A process of vapor barrier installation called "enveloping" entails the following:

a. __________ a. It seals entire walls with rolls of __________ film.

b. __________ b. It is applied over insulation with no vapor __________.

c. __________ c. It should be fitted tightly around openings and __________ if necessary.

d. __________ d. It should be applied over studs, plates, and window and door __________.

__________ 29. Application of vapor barrier to a house insures complete freedom from vapor leakage. (T or F)

To be used with Unit 47

Pages 610-618

Name ____________________

Score: (21 possible) ______

Study Unit 44

Sound Insulation

______ 1. Sound insulation has always been important in hotels. (T or F)

______ 2. The following appliances increase the noise level in a home *(one wrong)*: a. television; b. radio; c. hot water heater; d. stereo.

______ 3. A family room is sometimes known as an ______ living room.

4. Complete the following description of the way sound travels:

a. ______ a. Noises create sound ______.

b. ______ b. These radiate outward through the ______.

c. ______ c. They strike surfaces which ______.

______ 5. The construction of a home has a high degree of influence on sound transmission within the home. (T or F)

______ 6. Parts of a house such as a ceiling, wall, or floor are rated for their sound resistance by STC. These initials stand for Sound ______ Class.

______ 7. The higher the STC number, the better the insulation against sound. (T or F)

______ 8. Sound travels readily through *(one wrong)*: a. masonry; b. air; c. wall studs; d. electrical switches.

______ 9. The floor with the highest INR rating is *(one right)*: a. 0; b. –2; c. –5; d. +2.

______ 10. Two excellent examples of sound-absorbing materials are ______ tile and panels.

______ 11. Normal speech can be understood quite easily in a room with a sound transmission class of ______.

______ 12. A city has an average background noise level of ______ db.

______ 13. The STC rating of 8″ concrete block is ______.

______ 14. Sound that goes around poorly fitted doors follows what is called a ______ path.

______ 15. A 1″ square hole in a wall rated at STC 50 can reduce the wall's performance to STC ______.

______ 16. The sound insulation of double wall in detail D in Fig. 47-6 has an STC rating of ______.

(Continued on next page)

_________________________ 17. The STC rating of floor-ceiling combination of detail A in Fig. 47-7 is ____________.

_________________________ 18. A sound insulation product can be placed beneath a floor surface to reduce the transmission of ____________ sounds.

_________________________ 19. The STC rating of floor-ceiling combination shown in detail B of Fig. 47-8 is ____________.

To be used with Unit 48
Pages 619-632

Name ______________________

Score: (30 possible) ______

Study Unit 45

Plaster and Drywall

______ 1. Some good interior finish materials for walls and ceilings are *(one wrong)*: a. drywall; b. gypsum; c. pegboard; d. plaster.

______ 2. The base for a plaster finish is called ______.

______ 3. The types of plaster base used the most are *(one wrong)*: a. sheet metal; b. gypsum; c. wood stripping; d. insulating fiberboard.

______ 4. As plaster dries, ______ can develop around openings and corners.

______ 5. Plaster must be reinforced around openings and corners by the application of ______ metal lath.

______ 6. Pieces of wood located around openings to act as guides in a wall to be plastered are called plaster ______.

______ 7. Plaster is made of the following *(one wrong)*: a. plastic; b. sand; c. water; d. lime.

______ 8. Plaster is usually applied in ______ coats.

9. Match the coatings at the left to the descriptions at the right:

a. ______
b. ______
c. ______
d. ______
e. ______

a. sand-float
b. brown or leveling
c. putty
d. scratch
e. double-up

1. a finish coat containing lime and sand
2. the first plaster coat
3. a smooth finish coat containing no sand
4. a coat which combines both scratch and brown coats
5. a second plaster coat which is leveled during application

______ 10. A combination of scratch and brown plaster coats is used on ______ or insulating lath.

______ 11. A finish plaster coat that takes a good gloss paint or enamel finish is ______ finish.

______ 12. Such materials as fiberboard and plywood are called ______ wall.

______ 13. Drywall is always applied vertically to the wall. (T or F)

______ 14. Drywall is fastened to the studs with nails, screws, or ______.

______ 15. One advantage of drywall is that it can be applied without regard for conditions of moisture in framing members of the house. (T or F)

______ 16. Most drywall used as wall covering is ______ inch thick.

(Continued on next page)

_______________ 17. When drywall is nailed to the studs, begin at the outside and work toward the center. (T or F)

_______________ 18. When a nail head moves in the drywall after it is driven, it causes a nail _______________.

_______________ 19. The _______________ nailing system helps prevent movement between drywall and the nail head.

_______________ 20. It is best to cut drywall with a _______________.

_______________ 21. Joints between drywall should be smooth. This is done by applying joint _______________ and perforated tape.

_______________ 22. Joints must be sanded. (T or F)

_______________ 23. Drywall cannot be applied to ceilings. (T or F)

_______________ 24. Drywall is sheet material made up of gypsum filler faced with _______________.

_______________ 25. Drywall can be applied with *(one wrong)*: a. nails; b. cement; c. screws; d. adhesives.

_______________ 26. When the relative humidity is 50% and the temperature is 80 degrees F., the drying time for joint compound is _______________ hours.

To be used with Unit 49
Pages 633-650

Name ______________________

Score: (21 possible) ______

Study Unit 46
Wood Paneling

______________ 1. Paneling is popular for walls and ceilings because of the ____________ and warmth of wood.

______________ 2. Wood paneling can be installed directly over studs. (T or F)

______________ 3. The basic types of paneling are *(one wrong)*: a. plywood; b. particleboard; c. hardboard; d. solid wood strips.

______________ 4. Paneling comes in thicknesses from 1/4″ to ____________″.

______________ 5. Paneling should be delivered to the room in which it is to be installed ____________ hours before installation.

______________ 6. The sheet size of paneling is generally 4′ × 10′. (T or F)

______________ 7. Masonry walls should be waterproofed before the studding or furring is applied. (T or F)

______________ 8. Shims behind furring strips should be wood ____________.

______________ 9. A plywood panel fastened from the floor partway up a wall is called ____________.

______________ 10. A partial plywood panel cut to go partway up a wall is usually ____________ inches long.

______________ 11. A full-size plywood panel is ____________ inches long, and three panels can be cut from it.

______________ 12. Plywood panel can be purchased in a wide variety of styles and veneers. (T or F)

______________ 13. Plywood panel must be installed unfinished. (T or F)

______________ 14. When it is necessary to cut a hole in a piece of installed plywood such as an outlet box, do the cutting with a ____________ saw.

______________ 15. The same techniques for applying plywood to walls can be followed for fiberboard and hardboard. (T or F)

______________ 16. Wood paneling to be used for wall covering *(one wrong)*: a. comes finished or unfinished; b. should be seasoned; c. should be 10″ or wider; d. should have correct moisture content.

______________ 17. If wood paneling is installed horizontally *(one wrong)*: a. fewer pieces are needed; b. the process takes a shorter time; c. it makes the room appear longer; d. it makes the room appear higher.

______________ 18. The nail size for 1/4″ thick plywood should be ____________d.

(Continued on next page)

_______________ 19. When cutting openings for electrical outlets, a _______________ cut can be made with a saber saw to start the cut.

_______________ 20. A room that has a perimeter of 58′ will require _______________ panels.

_______________ 21. The perimeter of a room is calculated by adding the _______________ of the four walls.

To be used with Unit 50 Name ______________________

Pages 651-675 Score: (42 possible) ______

Study Unit 47

Wood and Vinyl Flooring

______ 1. A finish flooring of hardwood strips requires an underlayment. (T or F)

______ 2. The most popular hardwood floor is *(one right)*: a. maple; b. birch; c. oak; d. pecan.

______ 3. Hardwood strip flooring *(one wrong)*: a. is expensive; b. has beautiful grain; c. is different from any other floor of its kind; d. can be laid in a hexagon pattern.

______ 4. Strips for hardwood flooring *(one wrong)*: a. come in different thicknesses; b. come in different widths; c. are usually tongued and grooved to fit together; d. can be purchased in one uniform grade.

______ 5. The most popular thickness for wood strips is *(one right)*: a. 1″; b. 25/32″; c. 1/2″; d. 19/32″.

______ 6. The most popular width for wood strips is *(one right)*: a. 1 1/4″; b. 1″; c. 2″; d. 2 1/4″.

______ 7. Because flooring strips are made of hardwood, they come in random lengths. (T or F)

______ 8. Grading of hardwood strip flooring is supervised by the Department of the Interior. (T or F)

______ 9. There are ______ organizations which enforce grading practices.

______ 10. Red oak is graded higher than white oak. (T or F)

______ 11. Hardwood flooring is kiln-dried before delivery to the building site. (T or F)

12. Special care should be taken when storing hardwood flooring:

a. ______ a. The storage room should have ______ windows.

b. ______ b. The floor on which flooring is stored should be at least ______ inches above ground.

c. ______ c. The building should not be cold or ______.

d. ______ d. In winter the storage room should be heated to at least ______ degrees.

e. ______ e. Before installing, store flooring at the installation site at least four or five ______.

______ 13. Flooring should be installed after the following construction operations have been completed *(one wrong)*: a. plastering; b. electrical wiring; c. interior trim; d. plumbing.

______ 14. Flooring is ready for installation immediately after the bundles are unwrapped. (T or F)

(Continued on next page)

15. Before installation begins, a thorough inspection of the ______________ should be made.

16. The last step before actual installation is to cover the subfloor with asphalt-coated building ______________.

17. Studying the floor ______________ will help the carpenter decide where to start the finish floor installation.

18. When strip flooring will run from room to hall to another room, the installation should begin in the hall. (T or F)

19. When strip flooring is laid up to a room with a door and a different kind of flooring, the strip flooring should end under the center of the door when it is ______________.

20. The best way to lay strip flooring is *(one wrong)*: a. parallel to the length of the room; b. running continuously between adjoining rooms; c. with ends of strips aligned; d. crosswise of the room's length if room is sufficiently wide.

21. Nailing should be done as follows:

a ______________ a. If flooring strips are 25/32″ thick and 1 1/2″ or more wide, use 7d or ______________d nails.

b. ______________ b. Either ______________ nails or cut steel nails are satisfactory.

c. ______________ c. If flooring nails chosen are steel-wire flooring nails, they should be ______________d.

d. ______________ d. Steel-wire flooring nails should be coated with ______________.

22. Wood strip flooring cannot be laid in a house with a concrete slab. (T or F)

23. Strip flooring that has been surfaced at the factory requires no sanding after installation. (T or F)

24. Many traditional floor finishes have been replaced by ______________ finishes.

25. Safety instructions for applying floor finishes include plenty of ventilation and the use of a ______________.

26. Three types of urethane finishes include *(one wrong)*: a. oil-modified urethanes; b. moisture-cured urethanes; c. color-based urethanes; d. water-based urethanes.

27. Acid-curing finishes contain formaldehyde. (T or F)

28. Hardwood floor wax is available in ______________ or liquid form.

29. The wood preferred for plank flooring is ______________.

30. Pattern floors are also known as ______________ and design floors.

31. Strip flooring that is 3/8″ × 2″ requires ______________ board feet to cover an area of 100 square feet.

32. The most common type of resilient flooring is made of ______________.

33. An underlayment of plywood should be at least ______________″ thick.

34. Vinyl tile can be laid directly over wood flooring. (T or F)

35. Asphalt tile that is 9″ × 9″ square will require ______________ hours of labor if 100 square feet are to be installed.

To be used with Unit 51

Pages 676-684

Name ______________________

Score: (64 possible) ______

Study Unit 48

Ceramic Tile

______ 1. Ceramic tile can be used throughout the house to provide surfaces that are *(one wrong)*: a. durable; b. colorful; c. flexible; d. easy to clean.

______ 2. All ceramic tile is made from pure clay. (T or F)

______ 3. Tile is ______ at high temperatures to form a relatively hard material.

______ 4. All tile has a glazed face. (T or F)

______ 5. Nearly ______ million square feet of tile is installed each year in the United States.

______ 6. Custom lots of ceramic tile are made by hand. (T or F)

______ 7. Most ceramic tile is made in highly automated factories. (T or F)

______ 8. The body of tile is called a ______.

______ 9. Tile is fired in a ______ at high temperatures.

______ 10. The temperature and length of firing determines the water permeability of the finished material. (T or F)

______ 11. Highly permeable tiles are the least waterproof because they absorb more water. (T or F)

______ 12. The most permeable tile is called ______ tile.

______ 13. The least permeable tile is called ______ tile.

______ 14. Tiles that will be exposed to water should be ______ permeable.

______ 15. If a drop of water sits on top of the back of the tile that indicates the tile is towards the ______ side of the scale.

______ 16. Wall tile is generally a ______ tile with a relatively soft glaze.

______ 17. Wall tile is usually about ______ inch thick.

______ 18. Tile that is specially shaped to form a border around a tile installation is called ______ tile.

______ 19. Field tile is glazed on the top surface only. (T or F)

______ 20. Paver tile is at least ______ inch thick.

______ 21. Paver tile is intended for use on walls. (T or F)

______ 22. Handmade, unglazed pavers are known as ______ tile or ______ tile.

______

______ 23. Quarry tile is excellent for use on floors. (T or F)

______ 24. Mosaic tile is always ______ inches or smaller.

(Continued on next page)

_______________ 25. The lugs on lugged tiles automatically determine the proper spacing. (T or F)

_______________ 26. Epoxy adhesives were developed in the 1970s. (T or F)

_______________ 27. Adhesives should completely cover the back of the tile. (T or F)

_______________ 28. Adhesives are applied with a _______________ trowel.

_______________ 29. Traditional mortar for ceramic tile is made from *(one wrong)*: a. Portland cement; b. stone; c. sand; d. lime.

_______________ 30. Dry-set mortar consists of a mixture of Portland cement, sand, and certain additives. (T or F)

_______________ 31. Dry-set mortar mixed with latex or acrylic-modified liquids is weaker than dry-set mixed with water. (T or F)

_______________ 32. Epoxy dry-set mortars can be used on the following surfaces *(one wrong)*: a. plastic laminates; b. steel; c. copper; d. plywood.

_______________ 33. Mastic is an organic adhesive that comes ready-mixed. (T or F)

_______________ 34. Mastic is stronger than mortar. (T or F)

_______________ 35. Tiles are attached to a substrate with _______________ or mastic.

_______________ 36. The space between tiles must be filled with a grout. (T or F)

_______________ 37. Grout is a form of _______________.

_______________ 38. Grout comes in a wide array of colors. (T or F)

_______________ 39. Tile grout comes in two forms – plain and _______________.

_______________ 40. Any kind of waterproof sheet material that is used between the tile installation and the substrate is called a _______________.

_______________ 41. Tar paper is a common waterproofing material used beneath ceramic tile. (T or F)

_______________ 42. The abbreviation for chlorinated polyethylene is _______________.

a. ____________________

b. ____________________

c. ____________________

d. ____________________

e. ____________________

f. ____________________

g. ____________________

43. Identify the following tools:

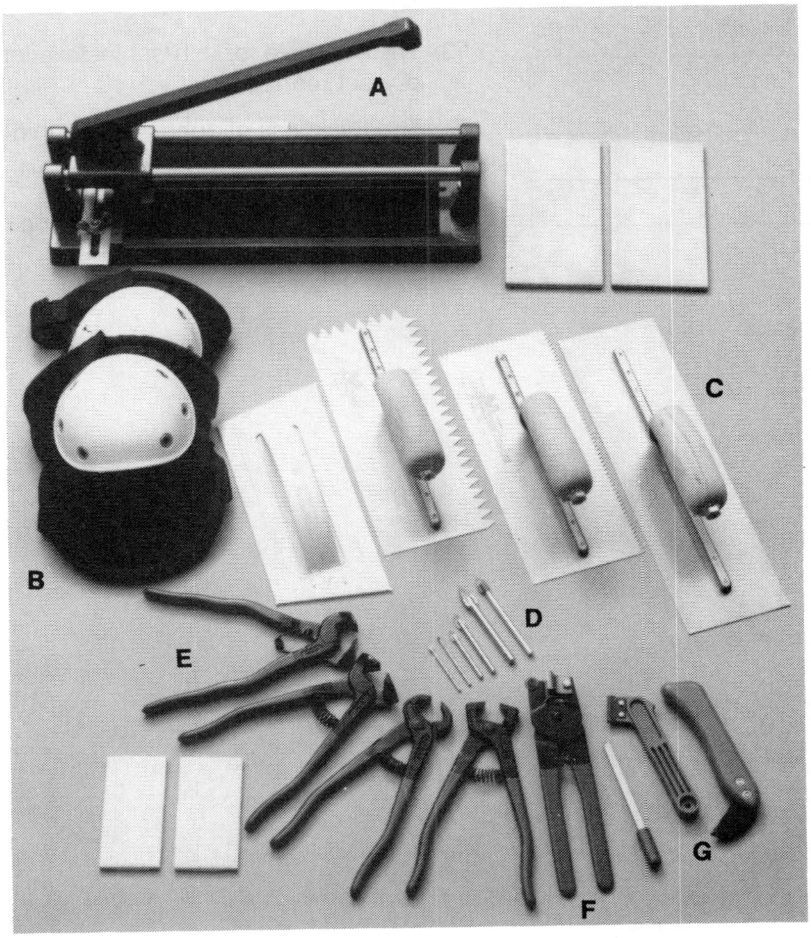

Fig. 48-1.

____________________ 44. A wet saw is a small radial-arm saw fitted with a __________ blade.

____________________ 45. There is only one method of installing tile. (T or F)

____________________ 46. The proper methods and materials used in installing tile depend on the following factors *(one wrong)*: a. structural loads expected; b. stiffness of the substrate; c. strength of tile; d. condition of substrate.

____________________ 47. One factor that can affect the stiffness of the floor is the distance between supports. (T or F)

____________________ 48. An isolation membrane reduces the chance that movement in the substrate will be transmitted to the tile. (T or F)

____

____________________ 49. Thick-bed installation requires a mortar-setting bed that is __________ inch or __________ inches.

____________________ 50. A thick-bed installation should be applied to metal lath. (T or F)

____________________ 51. The mortar for a thick-bed installation should be applied in __________ layers.

(Continued on next page)

______________________ 52. The layers include a scratch coat, a bed coat, and a ______________ coat.

______________________ 53. The adhesive for thin-set installations is made of ______________ or dry-set mortar.

______________________ 54. The thickness of adhesives for thin-set installations should be from ______________ inch to ______________ inch.
____

______________________ 55. A thin-set installation can be done over a concrete slab. (T or F)

To be used with Unit 52

Pages 685-706

Name ______________________

Score: (32 possible) ______

Study Unit 49

Stairs

1. The two main types of stairs are service and ________.
2. The ________ style is more expensive to construct.
3. A continuous-run stairway is straight. (T or F)
4. A stairway consisting of more than one run has a ________ or winder at the angle.
5. Of the two angle treatments, ________ is safer.
6. The part of a stair on which a person steps is called the ________.
7. Treads are supported on each side by ________.
8. Figure 49-1 shows a stairway with a ________.

Fig. 49-1.

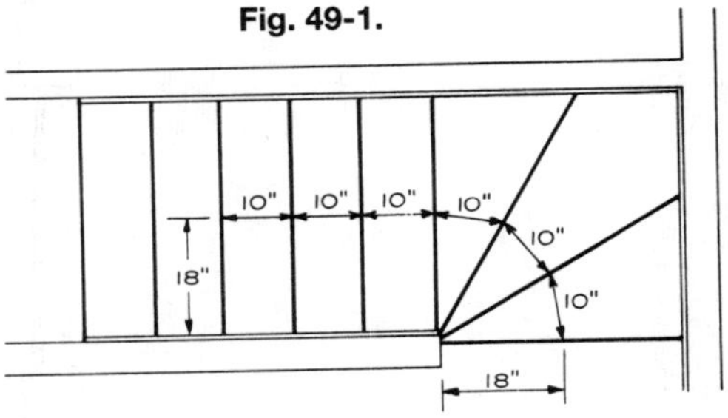

9. The section of the stairway from one step to the next is called a ________.
10. The edge of a step projects out a little beyond the vertical section. This is called ________.
11. The minimum amount of headroom required by FHA in designing a stairway is *(one right)*: a. 7′4″; b. 6′8″; c. 7′7″; d. 7′.
12. The full length of the vertical distance of a stairway is called the total ________.
13. The full length of the stairway measured horizontally is called the total ________.
14. The floor at the top of a stairway is called a ________.
15. A landing is the same thing as a platform. (T or F)
16. The opening for a stairway requires extra framing members. (T or F)
17. The most important designing considerations for stairways are *(one wrong)*: a. width of the stairwell; b. amount of headroom to be allowed; c. whether it is open or closed on one side; d. relationship between riser height and tread width.
18. The vertical distance between one floor and the ________ is total rise.
19. A stairway should be at least ________ feet wide.

(Continued on next page)

a. ____________________
b. ____________________
c. ____________________
d. ____________________
e. ____________________
f. ____________________
g. ____________________
h. ____________________

20. Name the parts of the stairs shown in Fig. 49-2.

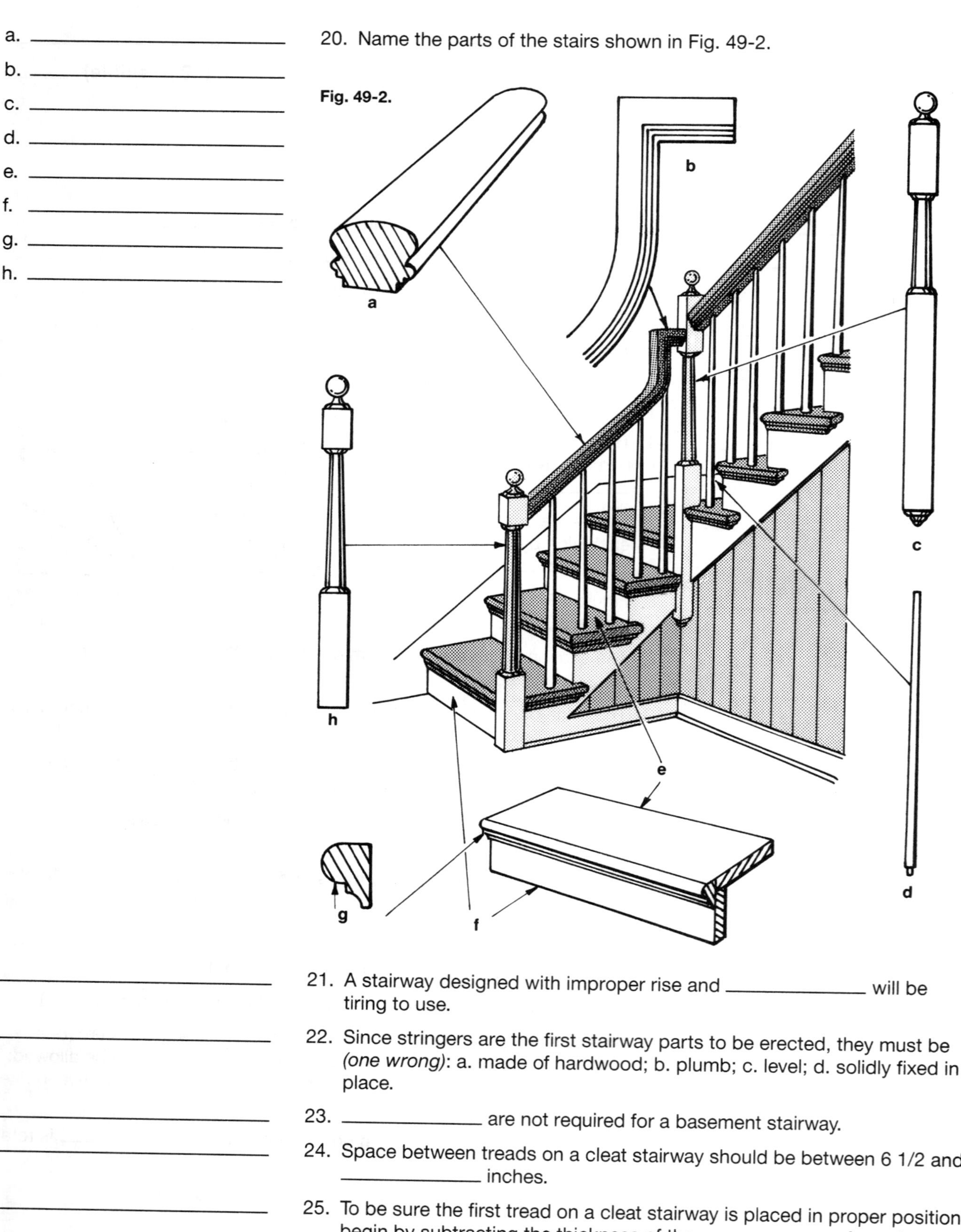

Fig. 49-2.

____________________ 21. A stairway designed with improper rise and ____________ will be tiring to use.

____________________ 22. Since stringers are the first stairway parts to be erected, they must be *(one wrong)*: a. made of hardwood; b. plumb; c. level; d. solidly fixed in place.

____________________ 23. ____________ are not required for a basement stairway.

____________________ 24. Space between treads on a cleat stairway should be between 6 1/2 and ____________ inches.

____________________ 25. To be sure the first tread on a cleat stairway is placed in proper position, begin by subtracting the thickness of the ____________ from the determined riser height.

To be used with Unit 53
Pages 707-738

Name ______________________

Score: (38 possible) ______

Study Unit 50
Cabinets and Built-Ins

a. ______________
b. ______________
c. ______________
d. ______________

1. Name the kinds of kitchen layouts shown in Fig. 50-1.

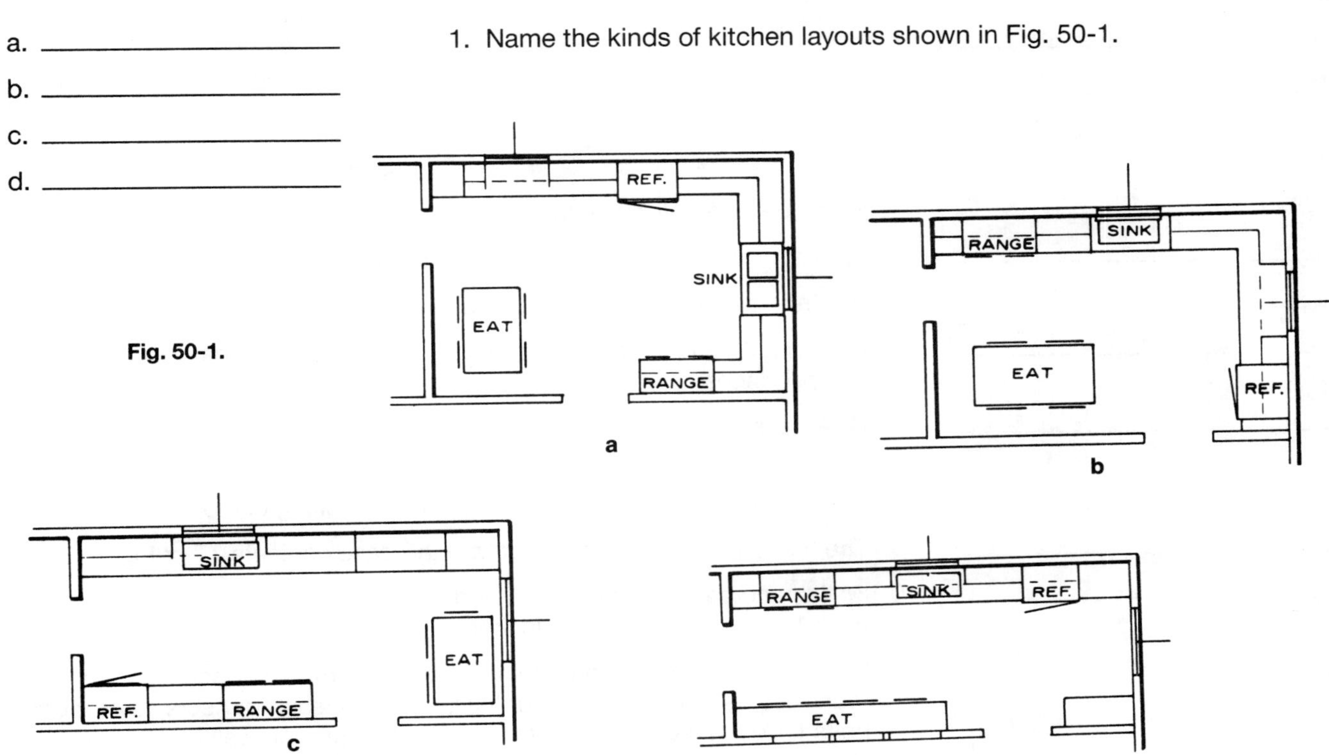

Fig. 50-1.

2. Complete these descriptions of kitchen layouts:

a. ______________ a. Working units are located along one wall in the ______________ design.

b. ______________ b. In the ______________ design, sink and stove are on one wall with the refrigerator on an adjoining wall.

c. ______________ c. Working units in the ______________ design are located on opposite walls.

d. ______________ d. The sink is on one end of the room and working units are on opposite walls adjoining the sink in the ______________ design.

______________ 3. The design usually found in small apartments is the ______________.

______________ 4. The main work centers of a kitchen are *(one wrong)*: a. food preparation; b. cooking; c. clean-up; d. dining.

______________ 5. Kitchen cabinets are produced in ______________ different ways.

______________ 6. When installing factory-built cabinets, measure along the wall ______________ inches from the floor.

(Continued on next page)

a. ______
b. ______
c. ______
d. ______
e. ______
f. ______

7. Match the dimensions at the left with the kitchen components at the right:

a. 24″
b. 60″
c. 12″
d. 21″
e. 36″
f. 30″

1. height of wall cabinets over a refrigerator
2. standard height of the counter of a base unit
3. width of the clean-up center
4. amount of counter space for serving center
5. height of wall cabinets over a sink
6. the usual height of a wall-hung cabinet

8. Electrical needs for a kitchen are:

a. ______ a. A source of ______ for each work center.

b. ______ b. General overall ______ for entire kitchen.

c. ______ c. Wiring that includes separate ______ for heavy-duty appliances.

d. ______ d. At least one double convenience ______ in each work center.

e. ______ e. A fan for proper ______ in the cooking center.

______ 9. The first cabinets to be installed are the base cabinets. (T or F)

______ 10. Wall space between base units and wall-hung cabinets should range from 14 to ______ inches.

______ 11. When installing wall-hung cabinets *(one wrong)*: a. choose #9 or #10 roundhead screws or toggle bolts; b. allow at least four screws or toggle bolts for each cabinet; c. fasten the screws through wall and 1″ into studs; d. align the tops of the wall cabinets 6′ from the floor.

______ 12. Kitchens usually consist of cabinets that are *(one wrong)*: a. built at the factory ready for installation; b. purchased in parts and sections ready for the carpenter to install on the job; c. built piece by piece, following the architect's plans; d. built individually to the carpenter's own specifications.

______ 13. Bathroom cabinets are similar in construction and installation to kitchen cabinets. (T or F)

______ 14. As a rule, bathroom cabinets are a little larger than kitchen cabinets. (T or F)

______ 15. Countertops are usually covered with plastic ______.

______ 16. The material most often used for countertop coverings can be described as *(one wrong)*: a. thin; b. requiring special tools; c. brittle; d. available in many sizes.

______ 17. Laminates must be fastened to a core ______ inch thick.

______ 18. Good cores include *(one wrong)*: a. solid wood; b. plywood; c. hardboard; d. particleboard.

______ 19. The core is fastened to the outer covering with ______.

_______________ 20. A drawer with one slide can be mounted in a frameless cabinet. (T or F)

_______________ 21. Double slides are available that allow the drawer to extend fully away from the cabinet. (T or F)

_______________ 22. European-style hinges are designed to be used on frameless cabinets. (T or F)

_______________ 23. The approximate time needed to install a base cabinet is _______________ hour.

To be used with Unit 54
Pages 739-776

Name ______________________

Score: (41 possible) ______

Study Unit 51

Interior Trim and Interior Doors

______ 1. The paint or natural finish to be used on interior trim determines the species of wood to select. (T or F)

______ 2. The moisture content for interior trim should be from 6 to ______ percent.

______ 3. Ponderosa pine would be a good wood to use when the trim is subject to hard usage. (T or F)

4. Before interior trim can be installed, the following construction must be concluded:

a. ______ a. The finish floor must be ______.

b. ______ b. The surfaces of walls and floors should be scraped ______.

c. ______ c. The location of all ______ should be clearly marked.

______ 5. Door and window frame trimming should be installed before base and wall moldings. (T or F)

______ 6. Openings in walls for insertion of interior doors are usually framed with two- or three-piece jambs. (T or F)

______ 7. It is possible to purchase a complete installation, including the jamb, with the door fitted and prehung. (T or F)

______ 8. The standard interior door thickness is *(one wrong)*: a. 1 1/4"; b. 1 1/2"; c. 1 3/8"; d. 1 3/4".

______ 9. Folding and sliding doors are usually thicker than standard interior doors. (T or F)

10. Identify the doors shown in Fig. 51-1.

a. ______

b. ______

c. ______

d. ______

Fig. 51-1.

a b c d

______ 11. The door frequently chosen for closets is the ______ door.

12. Match the interior door widths at the left with the door positions at the right:

a. 2'	1. closets
b. 2'6"	2. bathrooms
c. 2'4"	3. bedrooms
d. 6'	4. openings for sliding doors

a. ______

b. ______

c. ______

d. ______

(Continued on next page)

_______________ 13. Interior doors are a single standard height, regardless of their location in the house. (T or F)

_______________ 14. Standards for the direction of door swing are *(one wrong)*: a. against a blank wall; b. into a hallway; c. in the direction a person would naturally enter a room; d. so that the door doesn't interfere with the swing of another door.

_______________ 15. The hinges for an interior door should be loose-pin _______________ hinges.

_______________ 16. A door should always be hung with three hinges. (T or F)

_______________ 17. The door that allows greatest accessibility to a closet is *(one right)*: a. sliding; b. door that swings into the room; c. sliding bypass door; d. bifold door.

_______________ 18. A good choice for a place where the door is seldom closed is the sliding pocket door. (T or F)

19. Match the window trim parts at the left with their descriptions at the right:

a. _______________	a. casing	1. horizontal trim member that laps the window sill
b. _______________	b. sash stop	2. finish member below stool
c. _______________	c. stool	3. trim around the upper part of a window
d. _______________	d. apron	4. allows the sash to move freely

_______________ 20. Window trim is always made of wood. (T or F)

_______________ 21. The first piece of window trim to be installed is the _______________.

_______________ 22. The apron should extend the width of the _______________.

_______________ 23. The last trim to be installed in a room is the _______________ molding.

_______________ 24. Base molding can be purchased in a great many sizes and shapes. (T or F)

_______________ 25. Ceiling moldings are used *(one wrong)*: a. for decoration; b. as structural members which help support the ceiling; c. as junctions between wall and ceiling; d. to conceal any uneven appearance of paneling, drywall, or plaster.

_______________ 26. Wall moldings *(one wrong)*: a. can break up two different wall finishes; b. protect the wall from damage by chair backs; c. are installed from three to four feet high; d. are no longer used.

_______________ 27. Base moldings in a closet should match those in the room the closet serves. (T or F)

_______________ 28. The clothes pole for a closet should be about _______________ inches above the floor.

_______________ 29. The clothes pole fits into holes cut in two strips of molding fastened to the sides of the closet. (T or F)

_______________ 30. The closet shelf rests on top of the _______________ strips.

To be used with Unit 55

Pages 778-798

Name ______________________________

Score: (42 possible) ______

Study Unit 52

Chimneys and Fireplaces

______________ 1. A chimney must be of masonry construction. (T or F)

______________ 2. A chimney performs the following *(one wrong)*: a. produces a draft; b. helps supply fresh air to the fire; c. expels smoke and harmful gases; d. heats the house.

______________ 3. A chimney produces a better draft when outside temperatures are low. (T or F)

______________ 4. A chimney situated in the interior area of a house has a better draft than one on an outside wall. (T or F)

______________ 5. The gases and smoke go up a passage in the chimney called a ______________.

______________ 6. The two things most important to the efficiency of a chimney are flue size and ______________ height.

______________ 7. A chimney will not draw properly unless the topmost part of the chimney is *(one wrong)*: a. 2′ above a roof with a ridge; b. 3′ above a flat roof; c. even with any raised part of a roof within 10′ of the chimney; d. fitted with a hood to control irregular air currents.

______________ 8. A chimney is usually the heaviest single part of a house. (T or F)

______________ 9. Mortar for laying chimney flue brickwork should be made with ______________.

______________ 10. ______________ mortar does not resist heat and flue gases well.

______________ 11. A house with one heating unit and a fireplace requires ______________ flues.

______________ 12. A fireplace must have its own flue. (T or F)

______________ 13. A chimney flue ought to be lined because *(one wrong)*: a. the lining protects mortar and bricks in the chimney; b. the lining prevents cracks from developing in the chimney; c. this cuts the cost of construction; d. lined flues are safer and more efficient.

______________ 14. Flues should rise upward at no greater angle than *(one right)*: a. 40 degrees; b. 45 degrees; c. 30 degrees; d. 35 degrees.

______________ 15. An angle of ______________ degrees or less is the best rise for a flue.

______________ 16. Flues must always be built of solid masonry. (T or F)

______________ 17. Fireplaces, flues, chimneys and similar construction components must abide by the recommendations of an organization called ______________.

______________ 18. In a chimney containing more than one flue, 3 3/4 to 4 inches should be allowed for separating them. (T or F)

______________ 19. Fireplaces and stoves are connected to flues by smoke pipes. (T or F)

(Continued on next page)

__________ 20. Woodwork should never be closer than __________ inches to smoke pipes.

21. When a smoke pipe must pass through wood to reach a flue, one of the following precautions should be taken:

a. __________ a. Install at least 4″ of incombustible material or __________ around the pipe.

b. __________ b. Surround the pipe with a shield at least __________ inches larger than the pipe.

__________ 22. Between chimney walls and the wood members of the house, a __________ inch space should be allowed.

23. Complete the following statements on chimney-top construction:

a. __________ a. Moisture is prevented from entering between flue lining and brickwork by a concrete __________.

b. __________ b. The __________ lining extends 4″ above this.

c. __________ c. Rain and downdraft are kept out of the chimney by __________.

d. __________ d. The hazard of sparks can be controlled by means of spark __________.

__________ 24. The depth of a fireplace has little to do with good draft. (T or F)

__________ 25. Fireplace construction has been standardized for maximum safety and function. (T or F)

__________ 26. Floor framing around a fireplace must be stronger than in other areas. (T or F)

27. Match the fireplace parts at the left to the descriptions at the right:

a. __________	a. jambs	1. prevents downdraft
b. __________	b. hearth	2. the "floor" of the fireplace
c. __________	c. walls	3. exposed sides of fireplace
d. __________	d. throat	4. back and sides of fireplace
e. __________	e. lintels	5. extends from the top of the throat to the bottom of the flue
f. __________	f. smoke shelf	6. goes across the top of the opening to support the masonry
g. __________	g. damper	7. regulates throat opening
h. __________	h. smoke chamber	8. widest part of the flue

__________ 28. Modified fireplaces are made in a __________.

__________ 29. A modified fireplace *(one wrong)*: a. produces less heat than a conventional fireplace; b. contains all necessary fireplace parts; c. reduces chances of faulty construction; d. is less likely to smoke than a conventional fireplace.

__________ 30. A modified fireplace requires the construction of a conventional chimney. (T or F)

__________ 31. A complete fireplace plus chimney, called a __________ unit, can be purchased from the factory.

To be used with Unit 56

Pages 799-810

Name ______________________________

Score: (46 possible) ______

Study Unit 53

Protection against Decay and Insect Damage

______________ 1. Wood decay is caused by *(one wrong)*: a. extreme conditions of dryness; b. extreme conditions of wetness; c. water that works itself into wood parts of a house and soaks into the wood; d. decay organisms that grow in damp wood.

______________ 2. The two kinds of termites are dry-wood and ______________.

______________ 3. The kind of termite hardest to control is the ______________.

______________ 4. The termite that does the most serious kind of damage is the ______________.

______________ 5. Decay in wood can easily be seen on its surface. (T or F)

______________ 6. Wood decay develops most rapidly in temperatures of 70 to ______________ degrees F.

______________ 7. Wood decay organisms *(one wrong)*: a. are killed by kiln-drying of lumber; b. cannot grow in dry wood; c. cause wood to change in appearance and properties; d. are killed by cold temperatures.

______________ 8. Sapwood is more subject to decay than heartwood. (T or F)

______________ 9. Another name for wood decay is dry rot. (T or F)

10. Complete the following statements on prevention of decay:

a. ______________ a. Construction lumber should be fully seasoned, not ______________.

b. ______________ b. Wood should not be fully enclosed or ______________ until it is thoroughly dry.

c. ______________ c. Make sure there is proper clearance between wood construction members and the ______________.

d. ______________ d. Such units as steps, wall plates, and posts should be insulated with concrete or ______________.

e. ______________ e. Runoff of ______________ should be provided for on such sections as roofs.

f. ______________ f. ______________ should be installed around chimneys, doors, and windows.

g. ______________ g. A house should be fitted with ______________ and downspouts.

______________ 11. There are several kinds of ______________ that can be administered to wood to preserve it.

______________ 12. Preservatives can be administered *(one wrong)*: a. by pressure treatment; b. by painting; c. by soaking; d. by spraying.

(Continued on next page)

_______________ 13. Water vapor given off by indoor activities can cause wood decay unless the following precautions are taken *(one wrong)*: a. installation of proper vapor barrier on warm side of walls; b. good ventilation in attic spaces; c. leaking pipes are ignored; d. clothes washing done in basement.

_______________ 14. Using dry lumber in construction is the simplest way to prevent wood decay. (T or F)

_______________ 15. The termites that bore from the ground through tunnels into wood are the _______________ termites.

_______________ 16. _______________ termites fly directly to wood.

_______________ 17. Timber, woodwork, and furniture that are finished or painted are less likely to be attacked by termites than raw wood. (T or F)

_______________ 18. Protection against termites should be taken into consideration during both planning and _______________.

_______________ 19. Soil under the house should be kept as _______________ as possible.

_______________ 20. The best foundation for protection against ground termites is properly reinforced _______________.

_______________ 21. The best protection against ground termites is to treat the _______________ around and under foundations.

_______________ 22. Any wood used in decorative fences and gates should be pressure-treated with a good _______________.

23. The following precautions should be taken to prevent damage from dry-wood termites:

a. _______________ a. Give the exterior of the house several coats of _______________.

b. _______________ b. Fill all cracks and crevices with mastic _______________ or plastic wood.

c. _______________ c. Carefully inspect all _______________.

d. _______________ d. Put _______________ on all outside doors and windows.

e. _______________ e. Use construction lumber that has been given a _______________ treatment.

_______________ 24. When using pesticides, follow the label _______________ carefully.

_______________ 25. Carpenter ants eat wood. (T or F)

_______________ 26. To prevent carpenter ants, the chemical treatment must be applied to the _______________.

_______________ 27. The most common type of beetle is the _______________ beetle.

_______________ 28. The second type of beetle is the _______________ beetle.

_______________ 29. The greatest beetle activity is in wood with a moisture content between _______________ and 20 percent.

_______________ 30. All wood and wood products are subject to decay. (T or F)

_______________ 31. Dry-wood termites are found principally in *(one wrong)*: a. Florida; b. Maine; c. southern California; d. Hawaii.

_______________ 32. Fungi require air, warmth, _______________ and moisture to grow.

_______________ 33. Fungi grows most rapidly at temperatures of about 70 to _______________ degrees.

_______________ 34. Fungi cannot grow in wood with a moisture content of _______________ percent or less.

_______________ 35. Foundation walls should have a clearance of at least _______________" above the exterior grade.

_______________ 36. Water-repellent preservatives must have a minimum of _______________ percent by weight of pentacholorophenal.

To be used with Unit 57

Pages 811-818

Name ______________________________

Score: (27 possible) ______

Study Unit 54

Scheduling

______________ 1. The responsibilities of the general contractor include *(one wrong)*: a. arranging for materials needed by the subcontractors; b. job scheduling; c. material scheduling; d. coordinating construction work so it functions smoothly.

______________ 2. Successful general contractors usually are former ______________.

______________ 3. A general contractor should hire as few subcontractors as possible. (T or F)

______________ 4. Some of the jobs turned over to subcontractors include *(one wrong)*: a. plumbing; b. job scheduling; c. concrete and masonry; d. painting and decorating.

______________ 5. The delivery of needed materia s to the building site depends on *(one wrong)*: a. the date set for completion of the project; b. the size of the project and its type; c. the climate; d. the size of the work force.

______________ 6. The lumber supplier's responsibilities include *(one wrong)*: a. sending additional or missing items to the building site; b. accepting leftover materials for credit; c. keeping a financial record of the cost of the materials; d. collecting money from the contractor for payment for materials.

______________ 7. Filling earth around the outside of a foundation wall is called ______________.

______________ 8. Filling earth around the foundation cannot be done until and unless *(one wrong)*: a. the wood-frame walls of the building have been framed; b. the foundation walls have been braced; c. the framing has progressed far enough to provide sufficient weight to the foundation; d. exterior walls of the foundation have been moisture-proofed.

(Continued on next page)

a. ______________________
b. ______________________
c. ______________________
d. ______________________
e. ______________________
f. ______________________
g. ______________________
h. ______________________
i. ______________________
j. ______________________
k. ______________________
l. ______________________
m. ______________________
n. ______________________

9. Match the steps in home construction at the left with the workers who do them at the right. Some numbers will be used more than once:

a. installing pipelines in subsoil	1. carpenter
b. building chimney and/or fireplace	2. excavator
c. deciding when to backfill	3. plumber
d. finishing interior after plaster is dry	4. general contractor
e. planting lawns and shrubbery	5. painter
f. digging the basement	6. concrete worker
g. treating for termites	7. masonry worker
h. installation of doors and windows	8. landscaper
i. installing electrical wiring, switches, and fixtures	9. termite control specialist
j. framing walls, floors, and roof	10. electrician
k. putting the final finish on trim, doors, woodwork, and other similar jobs	
l. installing fixtures in bathroom	
m. doing the final cleanup	
n. putting in driveways and sidewalks	

______________________ 10. Interior trim is added immediately after walls are plastered or covered with drywall. (T or F)

______________________ 11. A building is inspected only once, in its final stages. (T or F)

______________________ 12. The contractor makes a ____________ list of things that the owner wants corrected.

______________________ 13. The easiest way to keep track of scheduling is with a ____________ chart.

______________________ 14. A scheduling method that shows the interrelationships between various tasks is called the ____________ ____________ method.

To be used with Unit 58 | Name ______________________________

Pages 819-837 | Score: (65 possible) ______

Study Unit 55

Plumbing, Electrical, and HVAC Systems

1. Plumbers must have skills in *(one wrong)*: a. woodworking; b. metalworking; c. painting; d. welding.
2. The letters used to identify the Uniform Plumbing Code are ______________.
3. Floor plans, elevations, and specifications for plumbing are developed by an ______________.
4. Residential plumbing consists of the ______________ supply system and the ______________ disposal system.
5. A water meter is always located inside the house. (T or F)
6. If a public water system is not available, the water must come from an underground ______________.
7. In a plumbing system, the ______________ must go through the roof.
8. Framing for bathtubs requires additional joists for support. (T or F)
9. Prefabricated showers and tubs are made of ______________.
10. Solder of lead and tin can be used to seal joints on copper tubing. (T or F)
11. The Drinking Water Act became law in ______________.
12. The letters used to identify polybutylene are ______________.
13. A nonlead solder is a combination of tin, copper, and ______________.
14. NCE stands for ______________ ______________ ______________.
15. An electrician must have a ______________ to install house wiring.
16. Identify these electrical symbols:

a. ____ a. S

b. ____ b. S_3

c. ____ c. ○ or -○-

d. ____ d. $○_{PS}$ or $-○-_{PS}$

e. ____ e. ⊖

f. ____ f. $⊖_R$

17. All power comes into the building through the ______________ entrance wires.

(Continued on next page)

_______________ 18. The following are the basic kinds of circuits *(one wrong)*: a. appliance; b. general purpose; c. general; d. special purpose.

_______________ 19. The simplest and least expensive wiring is called _______________ _______________ _______________.

_______________ 20. The bare copper wire in a circuit is used for _______________.

_______________ 21. Another name for nonmetallic sheathed cable is _______________.

_______________ 22. Another name for outlet boxes is _______________ boxes.

_______________ 23. All outlet boxes are made of metal. (T or F)

_______________ 24. There are two types of metal conduit, namely thin and _______________.

_______________ 25. Wiring is done in two stages, namely the _______________ and the finish.

_______________ 26. The electrical meter is always installed inside the house. (T or F)

_______________ 27. Inspectors must approve wiring at least _______________ times.

_______________ 28. Research is making heating and cooling systems more _______________-efficient.

_______________ 29. The four letters used to describe a heating, ventilating, and air-conditioning system are _______________.

_______________ 30. Heating, ventilation, and air-conditioning needs vary from region to region in the following ways *(one wrong)*: a. local conditions; b. climate; c. federal regulations; d. availability of resources.

_______________ 31. Electricity is more expensive in the Northwest than oil. (T or F)

_______________ 32. The Northwest gets much of its electricity from _______________ sources.

_______________ 33. With a forced hot-air system, a _______________ in the furnace circulates the warm air through ducts and registers.

_______________ 34. Forced hot-air systems respond slowly to outdoor temperature changes. (T or F)

_______________ 35. A hot-air system cannot be used in homes without basements. (T or F)

_______________ 36. A heat pump is a device that can only heat the air in a house. (T or F)

_______________ 37. A heat pump is connected to standard duct systems. (T or F)

_______________ 38. Filters should be used in a hot-air system. (T or F)

_______________ 39. Air cleaners that remove pollen, fine dust, and other irritants are _______________.

_______________ 40. A humidifier removes moisture from the air. (T or F)

_______________ 41. Ducts are made of sheet metal, rigid fiberglass, or flexible fiberglass _______________.

_______________ 42. Outlets for heating should be located high on the walls. (T or F)

_______________ 43. Warm air sinks towards the floor as it cools. (T or F)

_______________ 44. When a home has a basement, ducts are run above the ceiling joists and _______________ the floor joists.

_______________ 45. Homes without basements often have the ducts located within the slab. (T or F)

______________________ 46. Hot water and steam for a heating system are generated in a ______________.

______________________ 47. Two devices used to transfer heat into a room used with hot water or steam heat are radiators or ______________.

______________________ 48. Radiant heating systems can combine elements of hot-air and ______________ systems.

______________________ 49. In a radiant system, heat is transferred to a material that then ______________ the heat to people.

______________________ 50. With a radiant heating system, the heating coils or cables are buried within the ceiling, ______________, or walls.

______________________ 51. With a radiant hot-water system, heated water is circulated through tubing embedded in a ______________ floor.

______________________ 52. Heat recovery ventilators are sometimes called air-to-air ______________ exchanges or ______________ recovery ventilators.

____

______________________ 53. A heat recovery ventilation system usually is part of the regular heating system. (T or F)

______________________ 54. Whole-house ventilation can be accomplished with a large fan located in the highest ceiling in the house. (T or F)

______________________ 55. A two-pipe hydronic system uses one pipe to carry heated water to the room and one pipe to return the cooled water to the boiler. (T or F)

______________________ 56. Hot-air systems are designed to keep people comfortable, while radiant systems heat air. (T or F)

______________________ 57. HRV stands for heat recovery ______________.

______________________ 58. A device that can heat or cool the air in a house and that is connected to standard duct systems is known as a ______________ ______________.

To be used with Unit 59

Pages 838-858

Name ______________________

Score: (49 possible) ______

Study Unit 56

Interior and Exterior Painting

______ 1. Proper application of paint can prolong the life of a home and improve its ______.

2. Interior paints have the following purposes:

a. ______ a. They make surfaces easy to ______.

b. ______ b. They impart ______ resistance.

c. ______ c. They seal surfaces from ______.

d. ______ d. They enhance a desired ______ effect.

______ 3. The most popular interior paint is *(one right)*: a. high-gloss enamel; b. flat alkyd; c. latex; d. semigloss enamel.

______ 4. Drywall and plaster surfaces should be treated with sizing before painting. (T or F)

______ 5. The best paint for bathroom or kitchen is semigloss ______.

6. Interior wall finishing can be made a more pleasant job if the following pointers are kept in mind:

a. ______ a. Choose the best ______ for the job.

b. ______ b. Apply it according to the directions shown on the ______.

c. ______ c. Use ______ or rollers of high quality.

d. ______ d. Make sure walls are properly ______.

e. ______ e. Immediately wipe up all spills or ______.

f. ______ f. Make sure the room is ______ and well ventilated.

g. ______ g. Tools must be ______ immediately after use.

______ 7. Floors and furniture should be protected with newspaper or a ______ cloth.

______ 8. All cracks and nail ______ should be filled before painting.

______ 9. If a crack is large, fill it with ______ plaster.

______ 10. It is not necessary to allow patch repairs to dry before painting. (T or F)

______ 11. A professional paint job can be done without removing such hardware as switch plates and door knobs. (T or F)

______ 12. A narrow strip of paint should be applied around windows and doors and along ______.

______ 13. Applying a narrow strip of paint around the edge of a wall is called ______ in.

______ 14. One entire wall should be finished before going to the next. (T or F)

(Continued on next page)

_______________ 15. A roller is easier to use than a brush. (T or F)

_______________ 16. If a brush is used, the tip should be dipped one-half the length of the bristles. (T or F)

_______________ 17. Paint should be applied to a wall from the top downward. (T or F)

_______________ 18. Follow these techniques for painting woodwork *(one wrong)*: a. wait until walls are dry; b. use a wide brush; c. work as quickly as possible; d. avoid touch-up on areas that have dried.

_______________ 19. Painting window sash can be made easier by applying _______________ to the glass.

_______________ 20. There is no set procedure for painting doors. (T or F)

_______________ 21. Old paint can be removed by scraping or with a heat _______________.

_______________ 22. Protecting windows and other areas from being splattered with paint is called _______________ _______________.

_______________ 23. Interior paints can be applied with spray guns, _______________, and brushed.

_______________ 24. Exterior painting should be done before doors and windows are installed. (T or F)

_______________ 25. Oil-based paints are more popular than water-based paints. (T or F)

_______________ 26. Oil-based paints usually contain _______________ oxide.

_______________ 27. Paint protects the exterior of a home. (T or F)

_______________ 28. Paint's most important function is to keep excess _______________ out of the wood.

_______________ 29. Paint keeps exterior siding from cupping and warping. (T or F)

_______________ 30. Latex paint has the following characteristics *(one wrong)*: a. fast drying; b. easy to apply; c. can be applied to a damp surface; d. can be applied at any temperature.

_______________ 31. When applying a three-coat system, the first coat should be primer or _______________.

32. Some things to remember when painting are:

a. _______________ a. Seal knots and pitch spots with _______________.

b. _______________ b. Remove old scaling and _______________ paint.

c. _______________ c. Rough up slick, shiny surfaces with _______________ or a wire brush.

d. _______________ d. Sink any _______________ below the surface.

e. _______________ e. Clean all gutters and _______________.

f. _______________ f. Remove all loose or dry _______________ from around windows.

_______________ 33. Moisture condensation has a great effect on exterior paint surfaces. (T or F)

_______________ 34. Exterior walls must be protected against moisture from both inside and outside surfaces. (T or F)

_______________ 35. Outside wall paint is protected from moisture that comes from inside a home by a _______________ barrier.

To be used with Unit 60

Pages 860-865

Name ______________________________

Score: (23 possible) ______

Study Unit 57

Building Architectural Models

1. Building a model converts a drawing into a ____________-dimensional object. ______________________
2. Architectural models of homes are usually built to a scale of ____________″ = 1′. ______________________
3. Architectural models of large buildings are built to a scale of ____________″ = 1′. ______________________
4. Framing members of structural models should be made of walnut. (T or F) ______________________
5. There are two basic types of building models, namely, architectural and ____________. ______________________
6. The building model that shows primarily what the exterior of a building will be is the ____________ model. ______________________
7. A ____________ model is an actual construction of the building to scale. ______________________
8. A house model is best built to a scale of *(one right)*: a. one-half; b. one-fourth; c. one-eighth; d. one-sixteenth. ______________________
9. Woods used for a scale model should be hardwoods. (T or F) ______________________
10. The base of a model can be made of *(one wrong)*: a. hardboard; b. cement slab; c. plywood; d. particleboard. ______________________
11. It is best to assemble the model with regular-size nails. (T or F) ______________________
12. Building a model is merely an exercise in the use of tools. (T or F) ______________________
13. The roof of a model should be constructed as a single unit so it can be removed. (T or F) ______________________
14. Shrubbery for a model house can be made from ____________. ______________________
15. Lumber for structural members should be made of *(one wrong)*: a. basswood; b. ash; c. yellow poplar; d. redwood. ______________________
16. Attach shingles to the model with ____________″ nails or staples. ______________________
17. Model-size trusses can be constructed using a ____________. ______________________
18. Termite shields for a house should be made of thin aluminum or ____________ foil. ______________________
19. A model house without a basement can use a piece of ____________″ plywood to represent the concrete slab. ______________________

(Continued on next page)

_______________ 20. A beam made of layers of thin wood should be painted _______________ to represent a steel I-beam.

_______________ 21. Wood siding can be made from thin _______________ wood.

_______________ 22. A thin veneer can be used for roof _______________.

_______________ 23. It is easy to build miniature parts for millwork. (T or F)

To be used with Unit 61
Pages 866-873

Name ______________________

Score: (21 possible) ______

Study Unit 58
Construction Systems

______ 1. A wood pile is a round or rectangular post driven into the soil. (T or F)

______ 2. A technique known as ______ involves using a high-pressure stream of water to install a pipe.

______ 3. A pile driver is the best way of putting a pile into the ground. (T or F)

______ 4. Using 2 × 6s instead of 2 × 4s as wall studs permits a builder to install insulation that is ______ percent thicker.

______ 5. The mechanical fasteners shown in Fig. 58-1 are known as ______ clips.

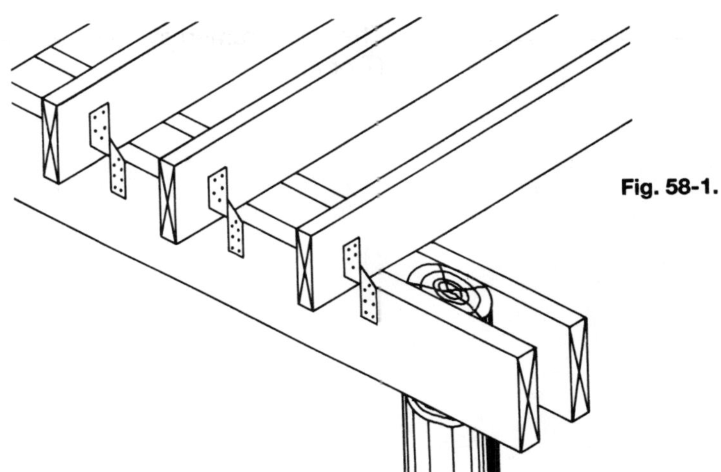

Fig. 58-1.

______ 6. Fire ______ block the flow of air upward inside walls.

______ 7. Materials are rated for fire resistance by the *(one right)*: a. American Association of Fire Chiefs; b. Underwriters' Laboratories; c. Associated General Contractors of America; d. United Brotherhood of Carpenters and Joiners.

______ 8. Using a mineral wood insulation in a wall decreases the wall's fire resistance. (T or F)

______ 9. Type-X gypsum board is designed specifically to resist sound. (T or F)

______ 10. Metal studs do not transmit sound as well as wood studs. (T or F)

______ 11. The permanent wood foundation *(one right)*: a. is difficult to install; b. can be installed only in good weather; c. can be built of untreated lumber and plywood; d. can be partially built in a factory.

(Continued on next page)

______ 12. Permanent wood foundation costs more than a comparable foundation of poured concrete or concrete block. (T or F)

______ 13. Fasteners used for a permanent wood foundation can be of plain steel. (T or F)

______ 14. Those areas of the earth that are subject to earthquake activity are known as ______ zones.

______ 15. In earthquake-resistant construction, each successive structural member above the foundation is tied to it. (T or F)

______ 16. Wood connectors are used to reinforce all structural joints. (T or F)

______ 17. The following sections of the country are subject to earthquakes *(one wrong)*: a. West Coast; b. Hawaii; c. Florida; d. East Coast.

______ 18. Truss frames are placed 16″ on center. (T or F)

______ 19. Truss frames are built on site. (T or F)

______ 20. Truss frames for an entire house can be erected in less than ______ days.

______ 21. The truss-framed system can be used only for single-story houses. (T or F)

To be used with Unit 62

Pages 874-898

Name ______________________

Score: (33 possible) ______

Study Unit 59

Remodeling and Renovation

________ 1. Paint peeling from random areas of a house's exterior is often caused by ________ from inside the house.

________ 2. A swayback or leaning roof ridge on an older home can indicate ________ settling.

________ 3. During the winter, wood window frames retain heat inside a house better than aluminum frames. (T or F)

________ 4. An attic should have at least ________ gable vent(s).

________ 5. The interior walls of most older homes are made of drywall. (T or F)

________ 6. An additional room often may be added to an existing house in the ________, the basement, or the garage.

7. Match each remodeling job to the percentage of its cost by which the value of the house increases:

a. ________	a. 25-30%	1. kitchen remodeling
b. ________	b. 50%	2. adding a full bath
c. ________	c. 80-100%	3. replacing windows and doors
d. ________	d. 75-80%	4. adding insulation

________ 8. The three steps in adding a room to an existing house are research, ________, and construction.

________ 9. The major difference between new construction and remodeling is the ________ process.

________ 10. Windows in older homes are held in place by nails driven into the exterior casing. (T or F)

________ 11. Some prehung doors are designed to fit into existing door frames. (T or F)

________ 12. Doors less than 7′ high need only two hinges. (T or F)

________ 13. Exterior steel doors are made with a(n) ________ core.

________ 14. All outside walls are bearing walls. (T or F)

________ 15. The simplest way to locate studs in a wall is with a stud finder. (T or F)

________ 16. Before cutting the opening for a new door in an existing wall, cut a(n) ________ opening.

________ 17. Partitions parallel to floor and ceiling joists are usually load-bearing. (T or F)

________ 18. A ________ is a prop placed against or beneath an object to support it.

(Continued on next page)

_______________ 19. A _______________ is a heavy wood beam placed across the top of several shores as a horizontal support.

_______________ 20. A leak in a roof will usually show up on the ceiling as a wet spot directly below the leak. (T or F)

_______________ 21. If a single wood shingle has a wide crack, it is best to *(one right)*: a. repair it using a sheet metal patch; b. replace the shingle; c. patch it with an asphalt shingle; d. cover the crack with roofing paper.

_______________ 22. Nails, rather than screws, should be used to straighten warped siding boards. (T or F)

_______________ 23. A suspended ceiling reduces noise in four ways. (T or F)

_______________ 24. A suspended ceiling absorbs a large amount of the noise striking its surface. (T or F)

_______________ 25. The grid system for a suspended ceiling should be at least _______________" to 2 1/2" below the bottom of the framing.

_______________ 26. In most cases, the main runners for a suspended ceiling should be installed parallel to the ceiling joists. (T or F)

_______________ 27. Tile should be cut with the face down. (T or F)

_______________ 28. Standard ceiling fixtures and _______________ may be used with a suspended ceiling.

_______________ 29. The panels used under ceiling lighting must be _______________.

_______________ 30. When installing ceiling panels, keep your fingers off the finished side of the board. (T or F)

To be used with Unit 63

Pages 899-909

Name ________________

Score: (28 possible) ______

Study Unit 60

Manufactured Housing

______ 1. Houses built stick by stick on a lot require more skilled personnel than factory-built structures. (T or F)

______ 2. Another name for factory-built housing is ______ housing.

______ 3. The first parts of houses to be prefabricated were *(one right)*: a. beams; b. trusses; c. flooring; d. roof girders.

______ 4. A method of house construction developed to promote industrialized housing is called the ______ Method.

______ 5. In the method of house construction developed by the National Lumber Manufacturers Association, the basic unit of measurement is *(one right)*: a. 4″; b. 6″; c. 12″; d. 16″.

______ 6. A prefabricated house consists of only the ______ of the house.

______ 7. About one-third of the total cost of a prefabricated house is for the ______.

8. A sectional home can be described as:

a. ______ a. Completely finished on the ______.

b. ______ b. Built on an ______ line.

c. ______ c. A complete home built in ______.

d. ______ d. Moved to the building site and then ______.

______ 9. A factory-built building unit designed to be used by itself or added to other units is a ______.

10. Match the manufactured housing at the left with the definitions at the right:

a. ______ a. recreational vehicle

b. ______ b. mobile home

c. ______ c. expandable mobile home

d. ______ d. sectional home

e. ______ e. double-width mobile home

f. ______ f. modular unit

1. made of two or more units joined to make a single house
2. a dwelling unit built for movement from one place to another
3. built on a chassis and meant to be used as a permanent dwelling
4. factory-built units for use by themselves or with similar units to make larger structures
5. a mobile home with parts that can be collapsed for transporting
6. a two-section mobile home, each with its own chassis

(Continued on next page)

________ 11. The maximum width of mobile homes is *(one right)*: a. 10′; b. 14′; c. 12′; d. limited by what each state will allow on the highway.

________ 12. The only manufactured housing that does not have to be moved on a truck is the recreational vehicle. (T or F)

________ 13. The most rapid growth in the manufactured housing industry is in the building of ________ homes.

________ 14. Workers in factories building mobile homes must be highly skilled. (T or F)

________ 15. Mobile homes are stronger in construction than modular housing. (T or F)

________ 16. A custom-built house is stronger than a modular home. (T or F)

________ 17. Factories producing modular houses require much heavy-duty equipment. (T or F)

________ 18. Modular homes are built of standard-size building materials. (T or F)

________ 19. The use of various types of assembly processes to manufacture houses is called ________ housing.

________ 20. Kitchens and bathrooms are the most complicated rooms in a house. (T or F)

NOTES

NOTES

NOTES

NOTES

NOTES

NOTES

NOTES

NOTES

NOTES